AF598606

International Series on
MICROPROCESSOR-BASED AND
INTELLIGENT SYSTEMS ENGINEERING

VOLUME 23

The titles published in this series are listed at the end of this volume.

An Introduction to Fuzzy Logic Applications

by

JOHN HARRIS

National University of Science & Technology, Zimbabwe

KLUWER ACADEMIC PUBLISHERS

DORDRECHT / BOSTON / LONDON

A C.I.P. Catalogue record for this book is available from the Library of Congress.

ISBN 0-7923-6325-6

Published by Kluwer Academic Publishers,
P.O. Box 17, 3300 AA Dordrecht, The Netherlands.

Sold and distributed in North, Central and South America
by Kluwer Academic Publishers,
101 Philip Drive, Norwell, MA 02061, U.S.A.

In all other countries, sold and distributed
by Kluwer Academic Publishers,
P.O. Box 322, 3300 AH Dordrecht, The Netherlands.

Printed on acid-free paper

Printed in the Netherlands.

To
William John Maxwell Duckworth

AN INTRODUCTION TO FUZZY LOGIC APPLICATIONS.

List of Contents.

AN INTRODUCTION TO FUZZY LOGIC APPLICATIONS

PREFACE

This text has its origins in the notes prepared for a number of short courses on the same topic, delivered to industrial and other professional groups. Although the principles of fuzzy logic have been known for more than three decades, they are not at present as widely understood or applied as their importance and value would justify. This is partly explained by the fact that most literate and numerate people are educated from an early age, whichever direction their education takes, to think in a classical logic way, that is to think in terms of true or false propositions. But fuzzy logic requires us to think in other terms than the quasi-classical logical, which is usually employed with the aid of various linguistic devices when the issues are not clear-cut. With fuzzy logic one is freed from this dependency, though it is natural to feel apprehensive in contemplating such a step because we are used, by long training, to continually striving to maintain intellectual rigour based upon classical logic principles. Relief is however found in recognising that classical logic appears as a special case, so it is not infact eclipsed by fuzzy logic, but for the real world it no longer occupies its pre-eminent position. To clarify the meaning, it is emphasised that it is not the logic itself that is fuzzy, but rather it is the information to which it is applied, the logic is actually multi-valued and therefore more flexible than classical logic, hence providing a language environment which is not antagonistic to fundamentally vague or imprecise information.

Through sufficient practise, articulation of problems in fuzzy logic terms becomes more familiar and quite natural. It is then recognised as a valuable tool in the expression and solution of cases where uncertainty, vagueness or ambiguity exist and where approximations are the norm. But now the approximations take on an entirely different character. Such cases abound in every profession and are the normal experience of everyone.

The text speaks to those engineers and others who may be interested in extending their ideas beyond the more usual deterministic treatments which pervade most current engineering education. These deterministic treatments are limited by their dependency upon a classical logic foundation. It is unlikely that a single text will suffice to cater for all needs and interests at the introductory level, but enough material is included to provide literacy in fuzzy logic. The methods are presented in symbiotic relationship with a cross-section of topics to found in current engineering education and point to the possibilities of wider integration into existing studies. There are many solved examples in the text to illustrate applications and to provide prototypes for further developments. The material presented is not exhaustive either in the methods illustrated or in the range of applications and it is hoped that the reader will feel encouraged to explore further than the boundaries of this text.

Physical systems, be they intelligent or not, all ultimately depend upon natural phenomena and the rules governing their behaviour. These rules are normally framed in classical logic form, though this is fundamentally incompatible with ambiguity or vagueness which is so often present to some degree in real problems. Part 1 of the text therefore illustrates fuzzy logic treatments of a representative selection of technological topics in processes and materials involving the processing of physical information. The resulting physical deductions in the form of fuzzy sets can be incorporated into knowledge based intelligent systems, as indicated in Chapter 1. Part 2 of the text takes up a systems perspective where physical phenomena are not explicit, though physical systems operations and organisation are all affected to some degree by fuzziness which has its origins in the underlying associated physical or psychological phenomena . In this part qualitative information and appropriate value judgements are incorporated, these are usually present in systems design, organisation and operations management.

It is possible that the reader will already have some acquaintence with the topics selected for fuzzy logic applications, as they often feature in whole or in part in many engineering and technology study programmes and it is partly with this in mind that they have been selected. For an introductory text it was deemed preferrable to present familiar types of material in unfamiliar terms rather than unfamiliar material in unfamiliar terms. Each chapter commences with introductory comments to recall some of the main features of the topic and to set the context for the following treatments. The text will be suited to engineering and technology courses, though some of the material covered later in the text may also be suited to business and commerce postgraduate programmes. Alternative fuzzy logic solutions to some problems are shown in the text and it is possible to compare the richer solutions obtained through the application of fuzzy logic with the current conventional treatments. If a deterministic model to a problems is available it is possible to fuzzify it to accommodate vague data, it is also possible construct a solution where no model is available provided only that an opinion is held about the nature of the system relation to a stimulus. The relationship may be refined by supervised or unsupervised learning using training data.

Programmes of study are subject to rapid development and the esoteric subject of today can become the commonplace of tomorrow. At the present time, the text would be suitable for special topics in related undergraduate courses in the engineering curriculum as well as for postgraduate courses Professional development short courses for the industrial, commercial, utilities, government and military sectors can also be based on material in the text.

There are relatively few texts available in this field at the present time in which problems are solved in fuzzy logic form, though technological applications are becoming increasingly available in the market for consumer products, especially where there is a control function. It will be readily appreciated that the field of application extends far beyond the material presented in this text, not only in the area of technology, but also in such divers areas as law, medicine, economics and sociology, all are likely to be increasingly influenced by the concepts and methods of fuzzy logic. This is surely just the beginning of what will ultimately become its widespread use in many fields, not only confined to technology.

Two books have been particularly useful in my studies: Fuzzy Logic with Engineering Applications by T J Ross (McGraw-Hill, 1995) and Fuzzy Controllers by L Riznic (Butterworth-Heinemann, 1997).

I am grateful for comments on the text by Dr.M.Biffin, Professor A.W.Lees and particularly Professor A.G.Atkins. But any omissions or errors in this text are mine and I shall be glad to receive reader's comments.

John Harris
1999

CHAPTER 1

COMMENTS AND DEFINITIONS

In science, engineering and many other areas of study, quantitative problems are solved using mathematical models which are cast in deterministic form, yielding closed form solutions or at least numerical solutions if this is not possible. This is the earliest form of science education and it is underpinned by classical (Aristotelian) logic, which has sets with sharp boundaries. In more advanced work the uncertainty inherent in some problems is treated through the application of statistical methods which account for random uncertainty through the concept of probability associated with multiple observations.

In many real cases, uncertainty is present in observations which is of a non-random nature. This is where, for example, in engineering a factor of safety would be introduced. The uncertainty may be due to the complex nature of the problem, such as the stress distribution in a component of complex geometry, or the flow pattern around a bluff body in a fluid with complex rheological properties. These are not uncertainties of a statistical nature and the corresponding concept is of possibility rather than probability.

It is this third category of problems where treatment on a basis of fuzzy logic (FL) is fruitful. As more knowledge about a system is accumulated and the uncertainty diminishes the need for an FL treatment also diminishes and it can revert to a deterministic or statistical one. This does however mean that in practice there is a wealth of valid FL applications and a freedom from the need to extend those based upon classical logic beyond their natural limits.

Some of the basic ideas of classical logic are recalled below. In this text * means algebraic product , but $\times$ means Cartesian product.(See the Appendix.)

1.1 Classical logic basics

The traditional classical logic (CL) defines distinct well-defined (sharp) sets, for example, the number of students registered for a course, or the names beginning with M in a given telephone directory. It also defines relations between sets called propositions. Consider for example, two sets: zebras and mammals, a simple proposition would be the assertion that all zebras are mammals, that is $Z \subset M$, where Z is the zebra set and M is the mammals set; $\subset$ means included in. The classical logic proposition is either true or false. (In this example it is true).

Elements (individuals) of a set are denoted by lower case letters, whilst capital letters denote whole sets: $n \in X$ means that n is a member of the set X.

A unit set contains one element denoted by (x). If a set is to be specified by listing all the elements, it is written thus: {a,b,c} in which the order of the elements is unimportant.

For example, the set of electrical light switch positions is S= {0,1}. The switch has two positions; 'on' or 'off'.

Special sets

i) Universal Set. This comprises all the elements under discussion, it is denoted by 1.

ii) Null or Empty Set. This set contains no elements, it is denoted by ∅.

Complementary sets

The complement of X is X`, where X`=1-X. (Also the complement of X` is X, where X=1-X`). It follows that the complement of the universal set is the null set ∅.

Logical operations

Useful CL operations are: 'and','or'.'And' is called the intersection ∩. 'Or' is called the union ∪.

These are illustrated by means of Venn diagrams, shown below in Figure 1.1.

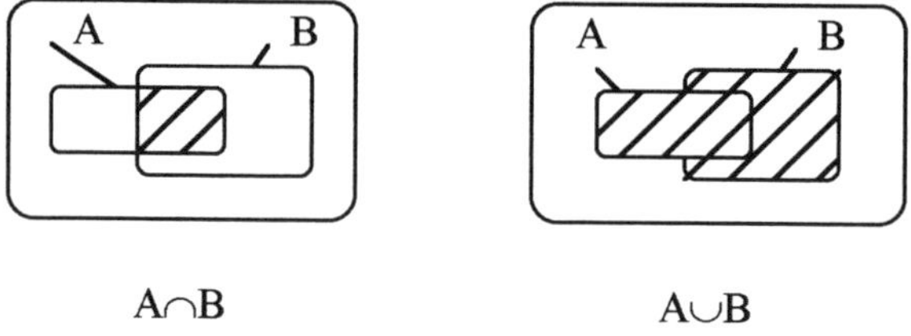

Figure 1.1 Venn Diagrams in Classical Logic

Venn diagrams are very useful as an aid to an intuitive approach to logic problems.The convention is that the surrounding rectangle of a diagram represents the universe of all the sets of the genre. Figures inscribed within the rectangle define arbitrary sets.

1.2 Fuzzy logic

1.2.1 Fuzzy Venn diagrams and sets

Sets in FL do not have sharp boundaries. A Venn diagram with two overlapping sets would appear as illustrated in Figure 1.2. The boundaries of the sets are not sharp as in the CL case, there is a degree of vagueness. A more useful way of illustrating FL sets is that shown in Figure 1.3 in which the membership (μ), $0 \leq \mu \leq 1$, is shown on the vertical axis for a continuous distribution of x.

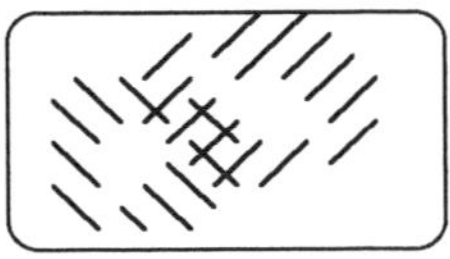

Figure 1.2 Fuzzy Venn Diagram. $A \cap B$

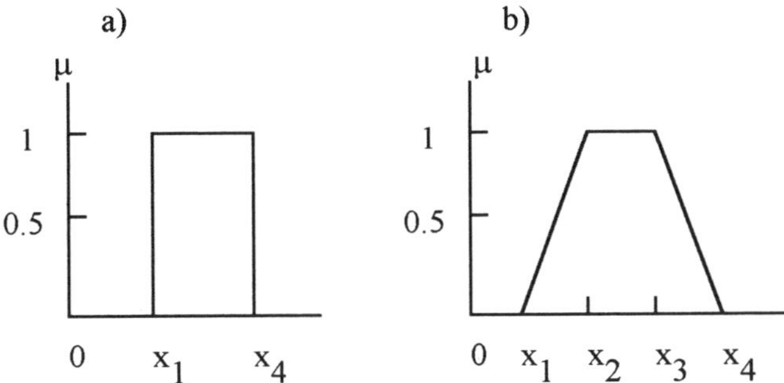

Figure 1.3 FL Set Membership

a) Classical set
b) Fuzzy set

Let the elements of the set be defined on the variable x and let μ represent the membership degree of the set. In case a) for the CL set, for $0 < x < x_1$ membership of the set is zero. For $x_1 < x$ the membership value is unity until $x = x_3$. In contrast to this for the Fl set, case b), for $0 < x < x_1$ membership is again zero,but for $x_1 < x < x_2$ the membership value gradually changes from zero to unity at $x = x_2$ .It remains at that value until $x = x_3$ when it commences to fall again (not necessarily at the same rate). Thus there is a gradual transition in this case from zero to full membership of the set and then back to zero again.

Example 1.1

Suppose that in Figure 1.3b), x represents temperature. Let $x_1 = 35$ C and $x_2 = 55$ C. Let the fuzzy set represent HOT.

Then $x < x < x_1$ represents NOT HOT
$x_1 < x < x_2$ represents HOT to grade μ
and $x_2 < x < x_3$ represents HOT

A temperature of 50 C represents HOT to grade 0.75.
A temperature of 40 C represents HOT to grade 0.25.

1.2.2 Universe of discourse

In the above example there is clearly a need to explain the treatment of a temperature above x_4 and below x_1. This can be done by increasing the number of linguistic terms to include, say, WARM and VERY HOT. The corresponding fuzzy set diagram is illustrated in Figure 1.4. It may be noted that there are now overlapping sets and that a temperature may have membership grade of more than one set. For example, temperature x has a membership grade of μ_1 WARM and μ_2 HOT, that is, it is predominantly HOT, but also WARM. It is not VERY HOT. Thus the need to make a categorical statement whether temperature x is HOT or WARM is avoided. This is achieved by partitioning the universe of discourse into several overlapping sets. The universe of discourse is the range of all the operating sets.

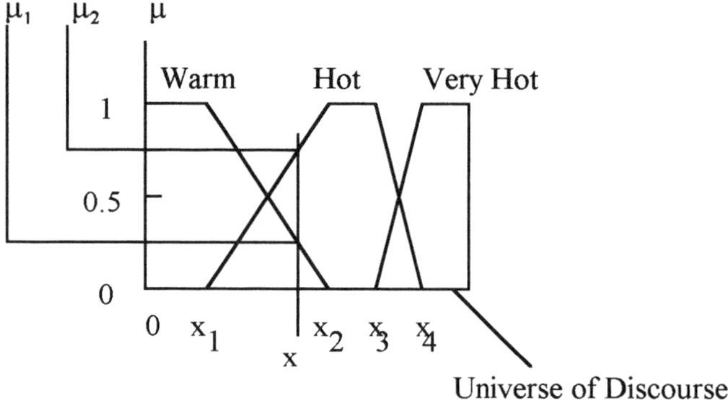

Figure 1.4 A Fuzzy Set Diagram for Three Sets

Thus in FL there is accommodation for uncertainty and it will be clearly seen that in this example it is not of a statistical nature.

1.2.3 Fuzzy set shapes

Fuzzy set shapes may be any convenient geometry dictated by exigencies of the problem at hand, and as indicated in Figure 1.4, they are not necessarily symmetrical. Typical shapes are illustrated in Figure 1.5.

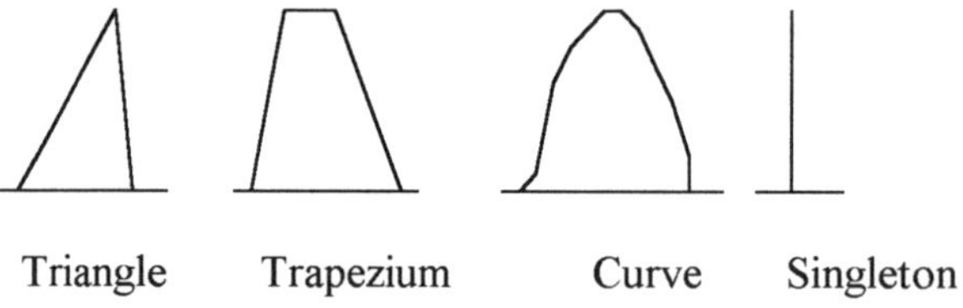

Figure 1.5 Typical Fuzzy Set Shapes

The singleton is clearly a limiting case of the triangular shape as the base length approaches zero. The most commonly used shapes are the triangle, trapezium and the singleton.

1.2.4 Propositions

Formal propositions are statements to be tested and in simple form comprise an antecedent (premise) and a conclusion. A CL proposition would be of the type; if today is Monday then yesterday was Sunday. CL asserts that either $\mu = 0$ or $\mu = 1$. Consider an FL proposition: if apples are ripe they are easy to pick. In this case there are various grades of ripeness between unripe and over ripe, and there are various grades of pickability. A (0,1) conclusion on the basis of ease of picking is not possible. But it will be seen in later chapters of this text how the problem might be formulated to reach an acceptable conclusion.

Compound propositions can take different forms. An important and often used form is: IF A AND B THEN C. For example; If students pass their examination and their industrial training then they graduate. Symbolically: $A \cap B = C$, where $\cap$ represents intersection. For the fuzzy logic case this may be illustrated as shown in Figure 1.6.

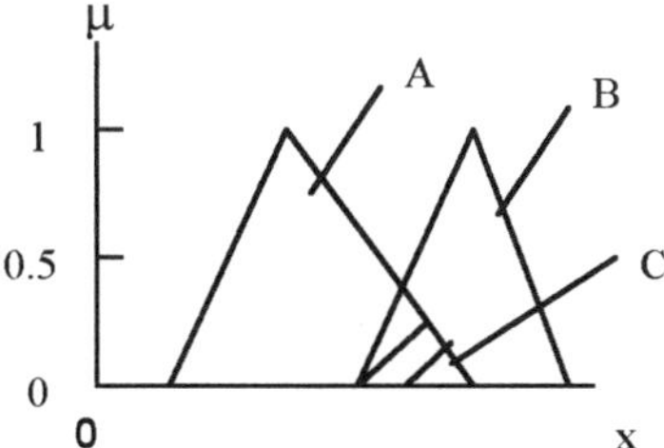

Figure 1.6 FL Intersection $A \cap B = C$.

This intersection is an example of a sub-normal (all $\mu<1$) convex set.

Another important and often used case of a compound proposition is expressed as: IF A OR B THEN C. For example: If learner drivers fail their theory or their practical test then they fail the driving test. Symbolically: $A \cup B = C$, where $\cup$ represents union. For the FL case this may be illustrated as shown in Figure 1.7.

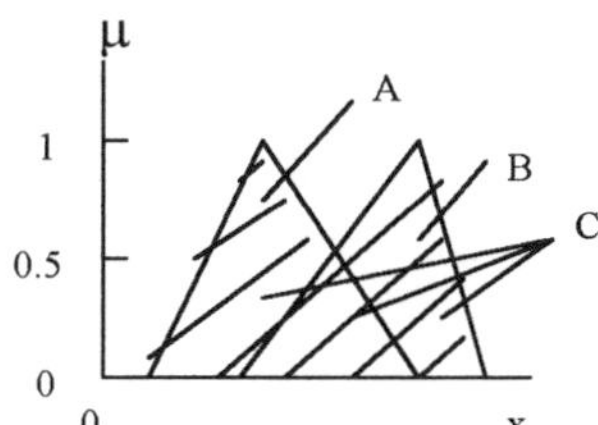

Figure 1.7 FL Union $A \cup B = C$.

The above union is an example of a normal non-convex set.

1.2.5 The complementary FL set.

If A is a fuzzy set, then the complementary set is given by, A` = 1-A. This illustrated in Figure 1.8.

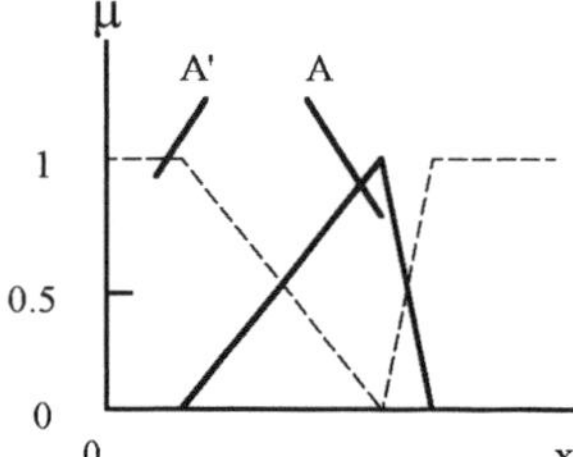

Figure 1.8 The Complementary Fuzzy Set A`.

The Appendix contains a guide to fuzzy logic operations used in this text.

1.3 Knowledge Engineering

The application of FL in practice may be viewed as a facet of a broad field of activity which is now called Knowledge Engineering. This is concerned with the organisation, operation and application of knowledge based systems. For example, use is frequently made of rule-bases and relational arrays (these are explained later), both of which are banks or archives of expert knowledge. The knowledge may have been captured from within the organisation or alternatively may have been purchased from external sources. A rule-base therefore represents embedded corporate knowledge and it is a resource which has a robust and enduring quality, less transient than the individuals within an organisation.

For the less industrialised regions of the world and also for small and medium sized enterprises (often called SMEs) in industrialised regions, rule-bases and relational arrays represent technology software that may be formally integrated into an organisation.

In the latter part of this text, FL applications in automatic control, industrial engineering and management operations are considered. The integration of a FL system into any operation needs to proceed in a rational way to ensure that the host is fully prepared. To avoid rejection there must be donor/receptor compatibility, as with any technology transfer. An outline of a formal procedure is given below.

1.3.1 Planning an integrated system

The overall plan for creating and implementing an FL knowledge based system should comprise four major steps as shown in Figure 1.9.

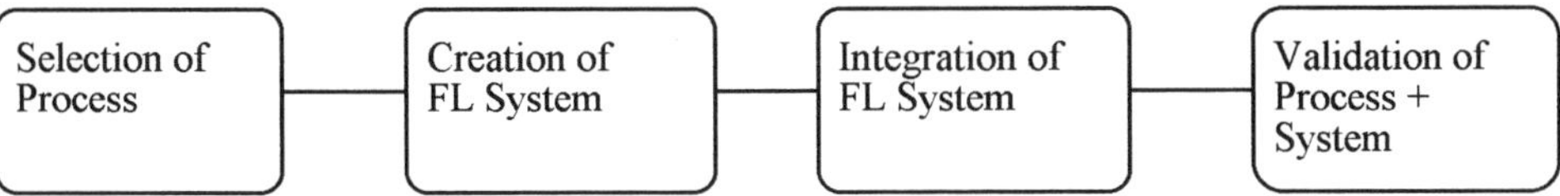

Figure 1.9 Steps for a FL Integration Process

In Figure 1.9 the process steps are not necessarily physical, they may represent steps in a service process; the output is then non-material. The steps are described in more detail below.

1.3.2 Process selection

The process needs to have an identifiable boundary and defined inputs and outputs. The essential features of these should not change in time. The process should have uncertain, vague, or ambiguous features which render CL inapplicable, but the uncertainty should not be of a random character that would make it a stochastic process.

1.3.3 Creating a FL system

This needs a planned approach and five steps are required as shown in Figure 1.10. The first step is identifying the parameters that constitute the antecedents and the conclusions also ranking and prioritising these, rationalising them where possible. The next step is identifying concepts and knowledge, which means finding or establishing conceptual models of the process and searching for existing process knowledge, also consulting expert opinion where local knowledge and concepts appears sparse. The third step is that of organising the knowledge into FL format, which means creating the appropriate propositions and the type of presentation.

Embodying in a FL framework includes; normalising the inputs and outputs, partitioning universes of discourse, fuzzifying and defuzzifying inputs and outputs, and also creating software or other information manipulation processes, such as look-up tables. Testing and validating means ensuring the stability and validity of the input. Pilot studies are also required to check the system with test data to ensure the acceptability of the range and quality of the results.

Figure 1.10 Creating a FL System

1.3.4 Integration of the FL system

The penultimate step is integration into the parent organisation. This may require some reorganisation and also retraining of staff. For example, in quality control the type of output will be different from the previous non-FL output and the functions of the staff will also change. The output may contain a judgement resulting from several inputs. Management as well as operator staff would require some adaptation to a FL based system, in the interpretation of the outputs and in the operation of the system.

1.3.5 Validation of the integrated system

Finally, to ensure a successful transition to the FL type of system a period of validation and commissioning is needed when the overall functioning and also the validity and quality of the results can be satisfactorily established under actual working conditions. This should be run in parallel with the existing system until the FL integrated system is established.

Various facets of the above will become apparent in reading the following pages though the actual integration into an industrial, commercial or other complex organisation is beyond the scope of the text. The above comments will however, offer guidelines on procedure when confronted with a new problem, even though it may not be the intention to formulate an organisational system.

1.4 Information and intelligence

Information is a coherent body of facts, such as a railway timetable, or a weather report. The information may be either quantitative (data) or qualitative. Intelligence is the ability to process or reason information, to learn and also to understand or know. In a system, which may be a corporation, the total system intelligence is the union of the human and artificial intelligence and it is an important resource. So too is the total system information which comprises that stored in the human brain, machine storage, audio-visual material and paper copies (text, drawings and images).

Knowledge can be considered as the intersection of information and intelligence. But whether machine reasoning and understanding exists is a philosophical question which is outside the scope of this text, it can however be deemed to exist in total corporate intelligence. Knowledge based systems are known as Expert Systems.

Reasoning can generally be divided into deductive reasoning and inductive reasoning. The deductive type is rule-following, whilst inductive reasoning is rule-making or generalising from particular instances, following recognised aims or objectives. It will be apparent from this text that rule-bases are key factors in the fuzzy logic method (or engine). A rule-base is a summary of a set of logic propositions with one or more antecedents and an inferred (or deduced) conclusion.

There are technical ways of constructing rule-bases and of partitioning universes of discourse by pattern recognition, they are also capable of being constructed from expert human knowledge. However they are constructed they must always be reconcilable with, and moderated by, human expert opinion.

1.5 Notation

Where the elements of a set are listed, these are written in lower case and are enclosed in brackets, {} thus,

$$A = \{a,b,c.........\} \tag{1.1}$$

Capital letters denote the whole set.

Where a discrete fuzzy set is expressed in terms of its elements and their associated membership values, they are enclosed in brackets [],

$$A = [\mu_1//a_1+\mu_2//a_2+\mu_3//a_3+.........]$$

The membership values μ_i are separated from the elements by the symbol //; this does not denote division. The + sign within the [] brackets implies continuation; it does not denote the algebraic sum.

In the solution of problems requiring the manipulation of fuzzy sets, the number of terms in the equation can sometimes become large. It is then convenient to define a principle set. The envelope of this discrete set contains all the elements of the complete fuzzy set. The envelope may be viewed as a piecewise continuous representation. Such a representation is frequently made use of in this text. The Appendix describes the method of deriving such principle sets.

CHAPTER 2

FLOW PROCESSES

This topic has its origins in the historical studies of hydraulics. It has since been widened to include studies on boundary layers, compressible flow, rarefied gases, plasmas and non-Newtonian flow. In all cases, simplifying assumptions are made to enable mathematical modelling to be achieved, though the resulting equations are sometimes rather intractable. The theoretical modelling was firstly in the form of inviscid flow theory, which was fruitful in solving flow fields at some distance from the solid boundaries. To this was added at a later date, Prandtl's boundary layer theory, thus bridging the gap between the distant flow field and the solid boundary. Theoretical, semi-theoretical and empirical treatments using dimensionless groups, model testing and computational methods has enabled satisfactory solutions to be found for many important cases. Expert opinion still plays a key role in evaluating results, because in real applications there are usually qualitative and some quantitative factors that are unmodelled.

As usual, CL is at the root of the theory and where two-valued logic falls short of reality, various factors are introduced to match theory and reality. FL methodology is a natural vehicle for the expression of problems and solutions where there is an element of uncertainty, imprecision or vagueness. Provided there is a general idea of the behaviour of the system under consideration and there is data and expert opinion available, then a FL framework can be constructed for analysis of the problem and presentation of conclusions. Examples of the value of the method are shown below for a case where there are unmodelled mechanical effects in an apparently simple hydrostatics problem and in non-Newtonian flow where there are frequent difficulties in modelling the rheological behaviour of the fluid.

2.1 Level control

For Newtonian fluids and also for non-Newtonian fluids with no yield stress, the pressure distributions determined by elementary analysis are entirely satisfactory provided that the fluid density and gravitational field strength distribution are both known. However, some substances do not exhibit a smooth flat free surface if disturbed, common examples being sludges and cement mixtures. The cause of this is that the material possesses a finite yield stress, even though it can flow without limit and has no memory of a reference state and is therefore a true fluid. In these substances the static stress distribution cannot be determined by elementary hydrostatic theory alone.

Even in the case of simple Newtonian fluids, a hydrostatic problem can be complicated by geometric and mechanical effects, including friction of interacting devices. In this case elementary principles may be inadequate. The following examples illustrates such a case expressed in several different ways.

Example Ex. 2.1

A storage tank is used as a buffer water supply and it is replenished from a mains water pipeline. The tank water level is controlled by a float-valve mechanism which has frictional effects so that the relationship between the water level and the valve opening is non-linear with hysteresis. The relationship is broadly governed by the fact that the valve is fully open when the water level is below 2.5m and is fully closed when it is above 3.5m.

It is required to
i) Create a FL solution for the system.
ii) Estimate the valve opening when the water level is 2.6m.

Solution

i) As a first step it is considered that three fuzzy linguistic sets would be sufficient to describe the universes of discourse of the water level and the valve opening as shown below in Table Ex 2.1 and Table Ex 2.2.

Table Ex 2.1 Partitioning of the Water Level H metres

		H					
	2.25	2.5	2.75	3.0	3.25	3.5	3.75
LO	1.0	1.0	0.5	0	0	0	0
ME	0	0	0.5	1.0	0.5	0	0
HI	0	0	0	0	0.5	1.0	1.0

Table Ex 2.2 Partitioning of the Valve Opening V%

		V%			
	0	25	50	75	100
CL	1.0	0.5	0	0	0
HO	0	0.5	1.0	0.5	0
FO	0	0	0	0.5	1.0

In the above tables the linguistic sets have the following meanings:

Water level: LO, ME, HI represent; Low, medium and high respectively.
Valve opening: CL, HO, FO represent; closed, half open and fully open respectively.

In general terms, the physical conditions of the system described indicate the following rule-base

Rule-Base:	**H**	LO	ME	HI
	V	FO	HO	CL

The corresponding causal relationship is contained in the following logic proposition

IF H THEN V

This completes the FL structure of the problem.

ii) If the water level is 2.6m the corresponding linguistic sets are obtained by interpolation in Table Ex.2.1. The result is 0.8 LO and 0.2 ME. Using the FL proposition and the rule-base, conclusions may be drawn as follows:

IF H	THEN V	MEMBERSHIP VALUE	CONCLUSION
LO	FO	0.8	0.8FO
ME	HO	0.2	0.2HO

The resulting valve opening is obtained from the union of the partial conclusions

$$V = 0.8FO \cup 0.2HO$$

This is illustrated as shown below in Figure Ex. 2.1.

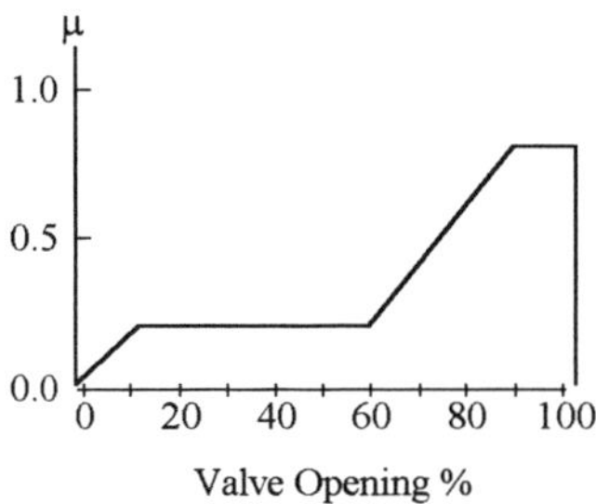

Figure Ex.2.1 Illustration of the Valve Opening Conclusion

The conclusion about the valve opening may be defuzzified giving 83% (discrete set), with a fuzzines metric of 0.1905 (see Appendix). But because of the lack of definition of hysteresis that is present in the system the single deterministic value obtained would not be very meaningful. The uncertainty in the valve position is best represented by a distributed set.

Alternative method

There are other methods of treating the same problem, for example, relational arrays may be created by using Cartesian products of fuzzy set pairs. The sets are paired using the rule-base and because there are three relations in the rule-base, there results three relational arrays. From Table Ex.2.1 and Table Ex.2.2 the fuzzy sets may be represented in discrete form as follows:

Water level: LO = 1.0//2.25+1.0//2.5+0.5//2.75+0//3.0+0//3.25+0//3.5+0//3.75
ME = 0//2.25+0//2.5+0.5//2.75+1.0//3.0+0.5//3.25+0//3.5+0//3.75
HI = 0//2.25+0//2.5+0//2.75//+0//3.0+0.5//3.25+1.0//3.5+1.0//3.75

Valve opening: CL = 1.0//0+0.5//25+0//50+0//75+0//100
HO = 0//0+0.5//25+1.0//50+0.5//75+0//100
FO = 0//0+0//25+0//50+0.5//75+1.0//100

The rule-base indicates that the following Cartesian products are to be formed: (LO,FO),(ME,HO) and (HI,CL), thus producing the following three arrays

		LO					
FO	1.0//2.25	1.0//2.5	0.5//2.75	0//3.0	0//3.25	0//3.5	0//3.75
0//0	0	0	0	0	0	0	0
0//25	0	0	0	0	0	0	0
0//50	0	0	0	0	0	0	0
0.5//75	0.5	0.5	0.5	0	0	0	0
1.0//100	1.0	0.5	0	0	0	0	0

HO	**ME** 0//2.25	0//2.5	0.5//2.75	1.0//3.0	0.5//3.25	0//3.5	0//3.75
0//0	0	0	0	0	0	0	0
0.5//25	0	0	0.5	0.5	0.5	0	0
1.0//50	0	0	0.5	1.0	0.5	0	0
0.5//75	0	0	0.5	1.0	0.5	0	0
0//100	0	0	0	0	0	0	0

CL	**HI** 0//2.25	0//2.5	0//2.75	0//3.0	0.5//3.25	1.0//3.5	1.0//3.75
1.0//0	0	0	0	0	0.5	1.0	1.0
0.5//25	0	0	0	0	0.5	0.5	0.5
0//50	0	0	0	0	0	0	0
0//75	0	0	0	0	0	0	0
0//100	0	0	0	0	0	0	0

By a fuzzy composition operation (see Appendix) defined by

$$V_i = R_i o[1.0//2.6]$$

where [1.0//2.6] is the fuzzy singleton representation of the water level of 2.6m, the following valve opening sets are obtained

Array	**V_i Set**
LO-FO	[0//0+0.5//75+0.8//100]
ME-HO	[0//0+0.2//25+0.2//50+0.2//75+0.8//100]

The values in these sets have been obtained by interpolation. The resultant fuzzy set is obtained by the union of the above components. Thus the resultant is

$$V = V_1 \cup V_2$$

(Noting that the HI-CL array yields a null set).

Hence, $V = [0//0+0.2//25+0.2//50+0.5//75+0.8//100]$

Comparing the above value of the valve opening with the fuzzy diagram, Figure Ex 2.1, it may be noted that the elements are in agreement, albeit with fewer points. The defuzzified value is 78%.

Second alternative method

For illustration purposes, a further alternative method is shown here, this alternative is through the application of the extension principle (see Appendix), which enables a fuzzy deterministic formulation to be created from the data supplied by assuming a fuzzy-linear model. The model selected is

$$V = K_1 - K_2 * H$$

where K_1 and K_2 are fuzzy constants. These constants may be solved by referring to the previously found R_i relational arrays. It is firstly convenient to find an overall relational array defined as the union of the three arrays found by the Cartesian products

$$R = R_1 \cup R_2 \cup R_3$$

The result of this operation is shown in Table Ex.2.3.

Table Ex.2.3. Overall Relational Array R

	H						
V	2.25	2.5	2.75	3.0	3.25	3.5	3.75
0	0	0	0	0	0.5	1.0	1.0
25	0	0	0.5	0.5	0.5	0.5	0.5
50	0	0	0.5	1.0	0.5	0	0
75	0.5	0.5	0.5	1.0	0.5	0	0
100	1.0	1.0	0.5	0	0	0	0

From the above array, if H = [1.0//2.5] then V = [0//0+0//25+0//50+0.5//75+1.0//100]

and if H = [1.0//3.5] then V = [1.0//0+0.5//25+0//50+0//75+0//100]

Inserting these values in the fuzzy linear-model above

$$[0//0+0//25+0//50+0.5//75+1.0//100] = K_1\text{-}K_2\text{*}[1.0//2.5]$$

and

$$[1.0//0+0.5//25+0//50+0//75+0//100] = K_1\text{-}K_2\text{*}[1.0//3.5]$$

by subtraction

$$[0//0+0//25+0//50+0.5//75+1.0//100]$$
$$-[1.0//0+0.5//25+0//50+0//75+0//100] = K_2\text{*}[1.0//1.0]$$

The fuzzy subtraction is shown below in tabular form

		V_1			
V_2	0//0	0//25	0//50	0.5//75	1.0//100
1.0//0	0//0	0//25	0//50	**0.5//75**	**1.0//100**
0.5//25	0//-25	0//0	0//25	**0.5//50**	**0.5//75**
0//50	0//-50	0//-25	**0//0**	**0.5//25**	0//50
0//75	0//-75	0//-50	0//-25	0//0	0//25
0//100	0//-100	0//-75	0//-50	0//-25	0//0

The principal elements of the fuzzy set are shown in the above table in bold type. The constant K_2 is therefore given by

$$K_2 = [0//0+0.5//25+0.5//50+0.5//75+1.0//100]$$

(The negative values have no physical meaning).

Now $K_1 = V+K_2\text{*}H$ (Algebraic sum)

As noted above, for H = [1.0//3.5];V = [1.0//0+0.5//25+0//50+0//75+0//100]

Therefore,

$$K_1 = [1.0//0+0.5//25+0//50+0//75+0//100]+$$
$$[0//0+0.5//25+0.5//50+0.5//75+1.0//100]\text{*}[1.0//3.5]$$

That is

$$K_1 = [1.0//0+0.5//25+0//50+0//75+0//100]+$$
$$[0//0+0.5//87.5+0.5//175+0.5//262.5+1.0//350]$$

The fuzzy addition is shown below in tabular form

	1.0//0	0.5//25	0//50	0//75	0//100	
0//0	**0//0**	0//25	0//50	0//75	0//100	
0.5//87.5		**0.5//87.5**	0.5//112.5	0//137.5	0//162.5	0//187.5
0.5//175	0.5//175	0.5//200	0//225	0//250	0//275	
0.5//262.5	0.5//262.5	0.5//287.5	0//312.5	0//337.5	0//362.5	
1.0//350	**1.0//350**	**0.5//375**	0//400	0//425	**0//450**	

The elements of the principal fuzzy set in the above table are shown in bold type. The fuzzy constant K_1 set is therefore

$$K_1 = [0//0+0.5//87.5+1.0//350+0.5//375+0//400]$$

The fuzzy-deterministic formula is therefore

$$V = [0//0+0.5//87.5+1.0//350+0.5//375+0//400]- [0//0+0.5//25+0.5//50+0.5//75+1.0//100]*H$$

Now considering again the given water level, H = [1.0//2.6]. The value of V is given by the following expression

$$V = [0//0+0.5//87.5+1.0//350+0.5//375+0//400]- [0//0+0.5//65+0.5//130+0.5//195+1.0//260]$$

The fuzzy subtraction is shown in the table below

	0	87.5	350	375	400
0	**0//0**	0//87.5	0//350	0//375	0//450
65	0//-65	**0.5//22.5**	0.5//285	0.5//310	0//385
130	0//-130	0.5//-42.5	0.5//220	0.5//245	0//320
195	0//-195	0.5//-107.5	0.5//155	0.5//180	0//255
260	0//-260	0.5//-172.5	**1.0//90**	0.5//115	0//140

(Negative values have no physical meaning). The physical values in the above table represent the percentage valve openings and therefore values greater than 100 are inadmissible.

The valve opening principal set is shown in bold type in the above table giving

$$V = [0//0+0.5//22.5+1.0//90+1.0//100]$$

There are only three terms in this expression and it is therefore more approximate than the previously found fuzzy sets for V. However, the general trend is the same, with the tendency towards the open state. The defuzzified value is 80.5%.

2.2 Non-Newtonian flow

Flow through round tubes is of primary engineering importance and consequently has been quite extensively studied. In non-Newtonian flow the prior work on Newtonian flow has frequently been used to provide guidelines because the results of other work should always converge to the accepted case as the fluid properties degenerate to those of a simple fluid of constant viscosity. Design data for Newtonian flow through round tubes is normally expressed as a correlation of friction factor with Reynolds number and displayed on a double logarithmic chart for flow through both rough and smooth tubes. This work is extensively covered in the literature. In functional form, the correlation is expressed as

$$F = f(R_e, \varepsilon) \quad 2.1)$$

where F is the friction factor defined by

$$F = 2\tau/\rho v^2 \quad 2.2)$$

and the Reynolds number is defined by

$$R_e = \rho v d/\eta \quad 2.3)$$

In the above expressions; v = mean fluid velocity, τ = wall shear stress, d = pipe diameter, ε = pipe surface roughness, r = fluid density and η = fluid viscosity.

The study of non-Newtonian flow is of much more recent origin, a substantial amount of the earlier work focused on ways of generalising the Newtonian treatment so that it could encompass non-Newtonian flow. But the achievement of this ideal has not been a simple matter. Under isothermal conditions and a moderate pressure regime, the rheological properties of a Newtonian fluid can, by definition, be described by a single physical quantity, namely the viscosity. Also by definition, a non-Newtonian fluid departs from this simple characterisation in one or more of the following ways:

i) The viscosity may be a function of shear rate.
ii) The viscosity may be a function of time under steady shearing conditions.
iii) The fluid may have a yield stress.
iv) Elastic effects may be present. That is, the fluid may have a fading memory of prior geometric configurations.

These various phenomena may all be present in varying degrees in one substance. This means that if any are present to a significant degree, then rheological characterisation by a single viscosity is no longer sufficient and that the Reynolds number, which is the ratio of inertia to viscous forces, will not generally be entirely adequate to characterise the state of flow. Concomitantly, it also makes the Reynolds number more difficult to define because there is no unique rheological property valid throughout the flow field.

Non-Newtonian flow behaviour is encountered in a wide variety of industrially important fluid classes; non-settling solid-liquid suspensions, liquid-liquid suspensions (emulsions), solutions of macromolecules and polymer melts are the most common. Common biological fluids such as blood and synovial fluid are also highly non-Newtonian. In some of these classes, the fluid inertia forces are relatively small (e.g. polymer melts), such classes are not further considered here. Settling solid-liquid suspensions are also industrially important, historically they have been treated on the basis of minimum transport velocity and are not considered as non-Newtonian fluids. From the above discussion it is apparent that the solution of problems in non-Newtonian flow is not as clear cut as in the Newtonian case, significant scatter is frequently found in the physical data and correlations for unsteady and turbulent flow may have substantial margins of error. Whilst some of the scatter may be treated on a statistical basis, non-random vagueness is more properly treated through multi-valued logic. Non-random vagueness or uncertainty will emerge in experimental data as trends which are not in accord with the predictions of theory. Such phenomena are not unusual in non-Newtonian flow. Steady laminar flow for both Newtonian and non-Newtonian fluids is usually amenable to treatment by deterministic formulae, and need not be considered here. There are two broad approaches to turbulent pipe flow design, one is based upon predictions using laboratory viscometric data, whilst the other is based upon a scale-up procedure from pilot plant data.

A well-known method of extending the usual Newtonian Reynolds number - friction factor correlation is outlined below. It essentially rests on using the expression for Newtonian viscosity in pipe flow to define an apparent viscosity in non-Newtonian flow. For Newtonian flow, the viscosity is

$$\eta = \tau_w/(8v/d) \qquad 2.4)$$

The term (8v/d) is the shear rate at the pipe wall in laminar flow for Newtonian flow, τ_w is the wall shear stress for any fluid.

For a non-Newtonian fluid with variable viscosity, the relationship between τ_w and (8v/d) is non-linear (and (8v/d) is no longer the shear rate at the tube wall). But the formula for the tangent to a point on the double logarithmic curve relating τ_w and (8v/d) may be expressed as

$$\tau_w = k(8v/d)^n \qquad 2.5)$$

An apparent viscosity may be defined by eliminating τ_w between equations 2.4) and 2.5) to yield

$$\eta' = k(8v/d)^{n-1} \qquad 2.6)$$

The viscosity in the definition of the Reynolds number is then replaced by the apparent viscosity given by equation 2.6) to yield an apparent Reynolds number defined by

$$R_a = \quad v^{2-n}d^n/8^n k \qquad 2.7)$$

Note that k and n in the above expression are not necessarily constant. If they are constant, then the fluid is called a power-law fluid. If not then they must be evaluated at the value of τ_w of interest.

It may be shown that the wall shear rate in laminar flow is given by

$$\gamma_w = ((3n+1)/4n)*(8v/d) \tag{2.8}$$

Hence a viscometric flow chart may be found from laboratory data using any convenient instrument. The chart may then be used to predict full-scale pipeline flow behaviour, even in the turbulent region. This procedure appears to be satisfactory for laminar pipe flow , but turbulent flow predictions are less satisfactory. One of the phenomena encountered is that turbulence can sometimes appear to be suppressed.

2.3 FL pipeflow design

In common with many industrial processes, the state of the art of non-Newtonian pipe flow design lies more in the realm of experience with specific fluid types rather than with general theory and deterministic equations. The FL structure of the problem of non-Newtonian pipe flow is treated below.

There are three system parameters (volumetric flow rate, pressure gradient and pipe diameter) and for a simple type of non-Newtonian behaviour, two fluid parameters (n,k). By using dimensionless groups these five parameters may be condensed to three (Reynolds number, friction factor and flow index n). These comprise two antecedents and a conclusion enabling a FL proposition of the usual form to be expressed

IF R_a AND N THEN F

This is a suitable form if the pressure gradient is the subject of the conclusion.

The universes of discourse and the partitioning of the physical parameters are selected with the following considerations:

i) Flow index N: Most non-Newtonian fluids likely to exceed laminar flow conditions would be of the so-called pseudo-plastic type, which means that the range of the flow index would be; $0<n<1$. This establishes the range of the universe of discourse. Equi-partitioning can be used because all values between about 0.2 and 1.0 are equally possible.
ii) Reynolds number R_a: The lower end of this would normally be about 2000. Although with some non-Newtonian fluids the divergence from laminar flow appears more gradual than for Newtonian fluids, the lower end is still the region of most rapid change. Finer partitioning is therefore required in this region. The logarithmic scale of the universe of discourse reflects the accuracy and definition of the range.
iii) Friction factor F: The greatest changes would be expected at the upper end of the range and therefore on the logarithmic scale the highest definition is sought through finer partitioning there.

The rule-base expresses the broad relationships between the linguistic sets of the parameters of the problem. An example is shown in Table 2.1 of an array of twenty five rules where the entries in the body of the array represent linguistic terms of the friction factor.

Table 2.1 A Friction Factor Rule-Base

	N		
$\mathbf{R_a}$	VN	MO	NN
SM	MH	ML	MH
MS	ML	MH	HI
ME	ML	MH	HI
ML	ML	ML	MH
LA	LO	LO	ML

The linguistic sets are defined as follows

Flow index:
VN = Very non-Newtonian.
MO = Moderately non-Newtonian.
NN = Near Newtonian.

Reynolds number
SM = Small
MS = Medium small.
ME = Medium.
ML = Medium large.
LA = Large.

Friction factor
LO = Low.
ML = Medium low.
MH = Medium high.
HI = High.

The rule-base is generally specific to the fluid type; suspensions would normally not be included in the same rule-base as solutions. The following example illustrates the FL method when the pressure gradient is unknown.

Example Ex.2.2

The friction factor is required for an aqueous solution of carboxymethyl cellulose, which is to be pumped through a pipeline at an estimated Reynolds number of 6000. Laboratory viscometric data indicate a flow index of 0.6. Assume that the partitioning of the Reynolds number, flow index and friction factor are as shown in Figure Ex.2.2 and that the rule-base in Table 2.1 is appropriate. Estimate the friction factor.

For this problem the Reynolds number, flow index and friction factor universes of discourse are partitioned as shown in Figures Ex.2.2, Ex.2.3. and Ex 2.4. respectively.

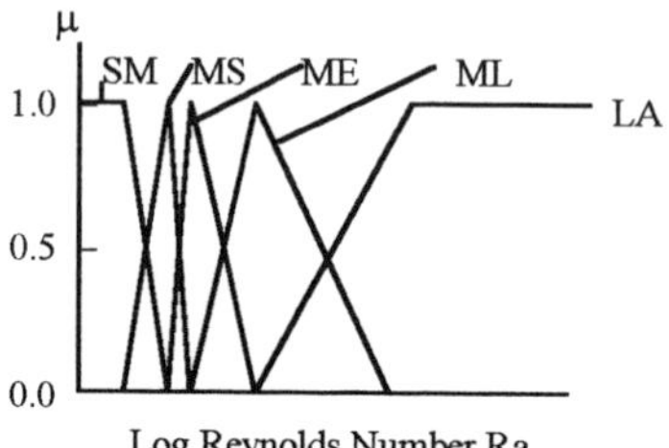

Figure Ex.2.2 Partitioning of the Reynolds Number
SM = Small MS = Medium Small ME = Medium
ML = Medium Large LA = Large.

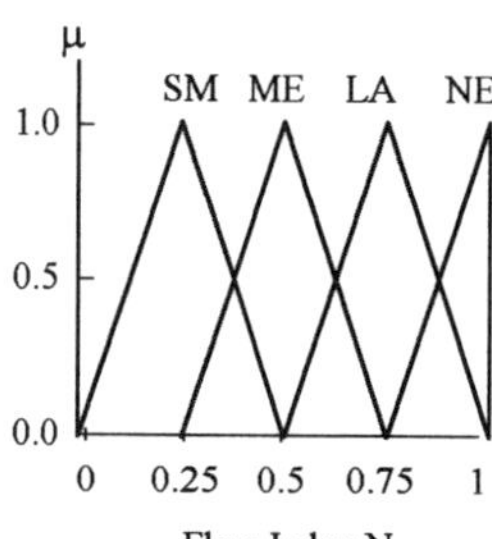

Figure Ex. 2.3 Partitioning of Flow Index
SM = Small ME = Medium
LA = Large NE = Newtonian

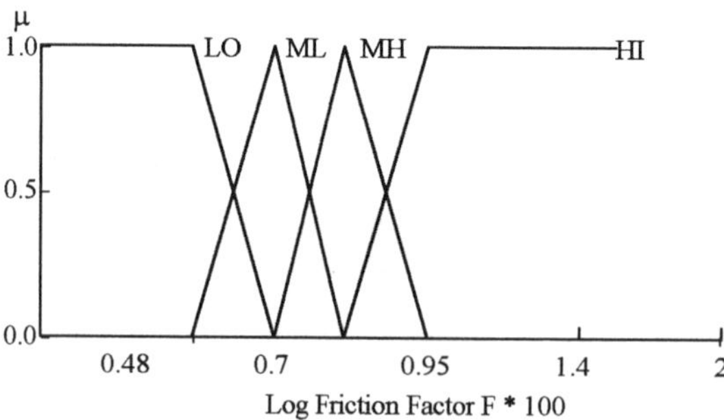

Figure Ex.2.4 Partitioning of the Friction Factor F
LO = LOW ML = Medium Low
MH = Medium High HI = High

Solution

With the given values of the Reynolds number and flow index the corresponding membership values of the linguistic sets are found from Figures Ex 2.2. and Ex 2.3 respectively.

$R_a = 6000$ $N = 0.6$
$\mu_{ML} = 0.75$ $\mu_{ME} = 0.43$
$\mu_{LA} = 0.25$ $\mu_{LA} = 0.57$

The FL proposition form and the rule-base may then be applied resulting in the following conclusions

IF R_a	AND N	THEN F	MIN	CONCLUSION
LA	LA	ML	0.25,0.57	0.25 ML
LA	ME	ML	0.25,0.43	0.25 ML
ML	LA	MH	0.75,0.57	0.57 MH
ML	ME	ML	0.75,0.43	0.43 ML

The result of the logic union of the above partial conclusions is shown in Figure Ex 2.5. below. Defuzzifying this result by the centroid method gives a friction factor value of 0.0074. For comparison, laboratory data for an aqueous solution (different material) flowing in relatively small tubes (1.27-5.08 mm diameter) with a flow index of n = 0.62 gives an average value of friction factor of 0.0066.

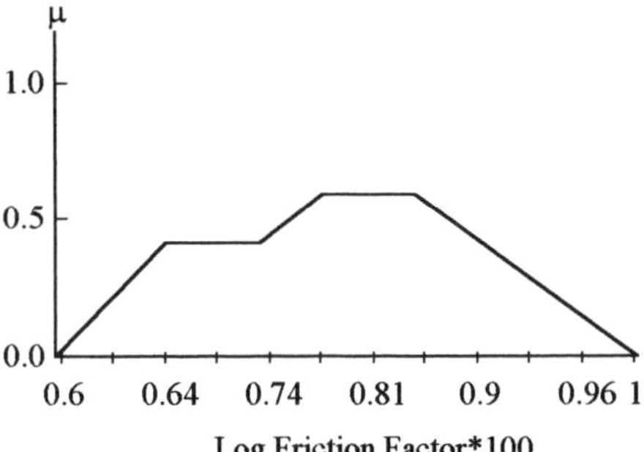

Figure Ex. 2.5 Fuzzy Friction Factor

Experimentally it will be found that with some fluids the flow index n is not constant over a wide range of shear rates. This makes the application of the usual friction factor-Reynolds number chart more difficult. But a spread of flow index numbers can be readily accommodated in the FL treatment. Such a case is illustrated for a simple triangular distribution of flow index in Figure 2.1.

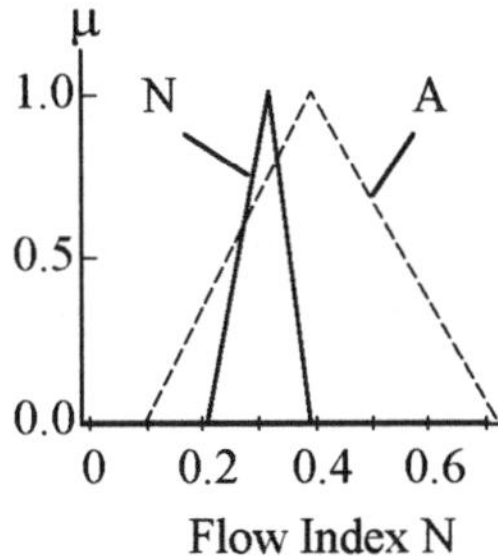

Figure 2.1 Distributed Fuzzy Flow Index
A = Partitioned Set. N = Particular Case

The maximum value of the fuzzy flow index intersection with a given linguistic set provides the membership value. There would usually be intersections with more than one linguistic set.

A similar case arises if the Reynolds number value is uncertain; a distributed value may be used as shown in Figure 2.2, again the intersection maxima are taken.

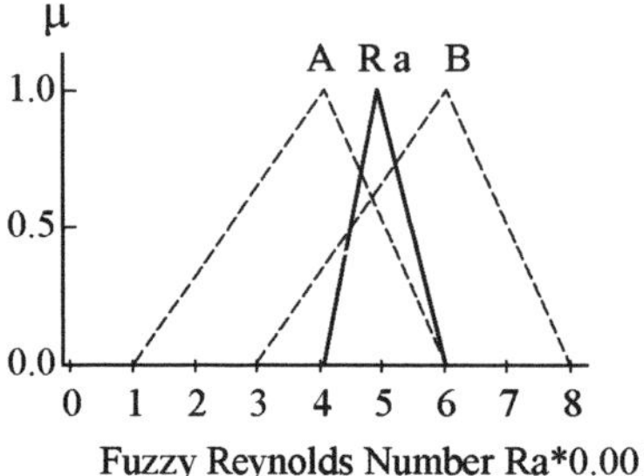

Figure 2.2. Distributed Reynolds Number

In Figure 2.2. A and B = Partitioned Sets. Ra = Particular Case.

Although the above method is able to accommodate variable viscosity, as found in many industrially important solutions and suspensions, it cannot cover gross under characterisation such as significant fluid elasticity or time dependency. If another phenomenon is to be included in the treatment then the logic proposition must be extended, for example

IF A AND B AND C THEN D

For long pipelines the capital and annual costs are often significant and a full study would account for the economic as well as the physical factors. An economic FL treatment would be conducted and integrated with the physical considerations. The Appendix describes the treatment of multiple antecedents.

2.4 Transport of solids

In the mining and processing industries pipeline conveying of solids is used as a convenient and cost-effective means of transporting particulate materials. Both pneumatic and hydraulic conveying are used, the former generally for finer particles and the latter for larger particle sizes of materials which can be wetted. Hydraulic conveying overlaps with non-Newtonian flow studies, the difference being that non-Newtonian fluids are assumed to be homogeneous, whereas in hydraulic conveying the disperse phase may (and often does) have an inhomogeneous concentration profile across the pipe diameter, including in extreme cases a bed of solids on the lower internal wall of the pipeline which ultimately may lead to blockage.

In practice, solids transport has usually been treated on a different theoretical basis to non-Newtonian flow. Because of the phenomenon of settling of the disperse phase there is the additional concept of minimum transport velocity, which expresses the lowest safe superficial axial velocity of the flow to avoid blockage of the pipeline. In hydraulic and pneumatic conveying it is usually assumed that non-Newtonian effects

are relatively unimportant, this means that the volume concentration of fines (< 30 microns effective diameter) is small.
Long pipelines require substantial capital investment and they therefore justify an experimental programme to establish the technical data for full-scale design. Short pipelines however do not warrant sufficient capital investment for a preliminary programme of work and therefore recourse must be made to prior knowledge and experience for guidance on optimum design. In short pipelines the entry length required to accelerate the disperse phase and achieve a uniform flow field is usually not a negligible fraction of the total pipe length. It introduces an additional pressure drop above that required to maintain steady flow and may infact represent a major part of the overall pressure drop.

A prominent role in the conveying of solids is played by the Froude number, defined below. This is a measure of the ratio of inertia to gravitational forces of the disperse phase. It is important because of the tendency of the disperse phase to settle, in non-Newtonian flow studies the Froude number is very small.

Friction effects in fully developed flow

Pumping power requirements are increased by the presence of the dispersed phase and this increase has been the focus attention. Pumping power is absorbed by the fluid in overcoming friction, supplying elevation of the fluid where needed as well as kinetic energy to the stream. Increasing the solids concentration in the fluid increases the friction which manifests itself as an increased pressure gradient for a given superficial velocity, this is true whether the continuous phase is water or air. the effect is typically as shown in Figure 2.3.

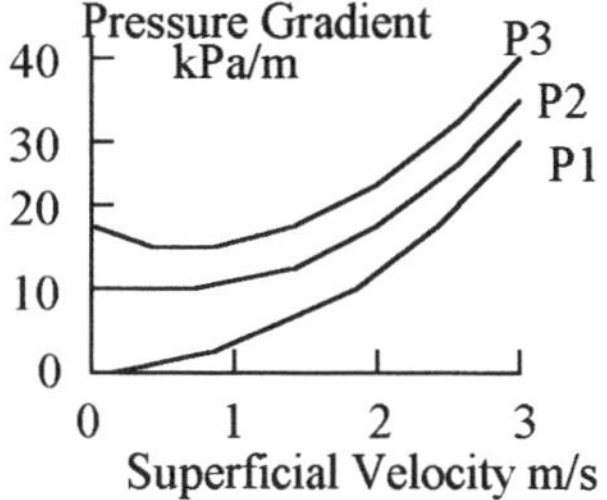

Figure 2.3 Effect of Solids Concentration on Pressure Gradient
0.3 m Diameter Pipe. P1-Low Concentration, P2-Medium Concentration, P3-High Concentration.

For settling systems, gravitational effects are clearly important, the available experimental data indicates that the friction factor of the two-phase fluid is a function of the Reynolds number(R), the particle Froude number(P), the fluid stream Froude number(S) and the volume concentration of solids(c). Thus the friction factor is given in functional form by

$$F = f(P,S,R,c) \qquad 2.9)$$

Alternatively, since the fluid Reynolds number may be expressed as a function of the fluid friction factor in the absence of solids (F')

$$F = f(P,S,F',c) \quad 2.10)$$

In equations 2.9) and 2.10), the particle Froude number is defined by

$$P = u^2/gd\delta\rho \quad 2.11)$$

where u is the settling velocity, d is the effective particle diameter and $\delta\rho$ is the fractional density difference between the dispersed and the continuous phases

$$\delta\rho = (\rho'-\rho)/\rho \quad 2.12)$$

The published data is sometimes cast in the form of a friction factor difference, δF

$$\delta F = (F-F')/F' = cf(P,S) \quad 2.13)$$

The hydraulic gradient difference is also frequently used as an alternative, this is defined by

$$\delta i = (i-i')/i' = (F-F')/F' \quad 2.14)$$

It will be noted in the above that F' (or i') must be evaluated under identical conditions of flow and temperature and in the same type of pipeline as F (and i).

The particle drag coefficient, K, is related to the particle Froude number by

$$K = 4/3P \quad 2.15)$$

and therefore equation 2.13) may be recast in the form

$$\delta F = cf(P,K) \quad 2.16)$$

The type of relationship found for experimental data is illustrated in Figure 2.4.

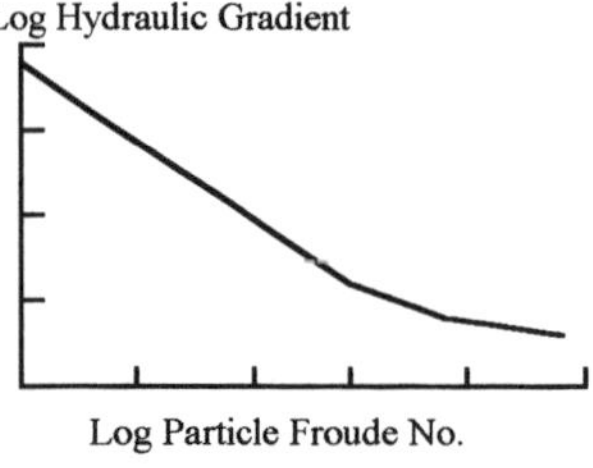

Figure 2.4 Typical Effect of the Particle Froude Number on the Hydraulic Gradient

It is also possible that the friction factor might also depend upon the ratio of the particle settling time to the process time scale but this is not usually considered.

It may be noted from Figure 2.4, for the lower range of Froude number, that approximately

$$iP^{w} = \text{constant} \qquad 2.17)$$

where w is a positive constant. It is also found experimentally that the trend represented in equation 2.18) is approximately true

$$\delta FS^{-x} = \text{constant} \qquad 2.18)$$

where the index x is another constant.

There is significant scatter in the data for both the above expressions. The rule-base shown in Table 2.2 reflects the trends expressed in equations 2.17 and 2.18), but in contrast the relations are not deterministic, which is a more realistic representation of the level of vagueness in the state of knowledge. In practice, suspensions have a range of particle sizes and shapes and the distribution of these would have an effect on the gross properties of the suspension.

Example 2.3

A suspension of particles having a mean effective diameter of 2 mm is to be pumped through a 76.25 mm diameter pipe which is 300 m long. The volumetric flow rate of the suspension is 0.0139 m^3/s. The specific gravity of the solid particles is 4.0 and the density of the suspension is 1179 kg/m^3. The average particle settling velocity is 0.108 m/s. Assume that the pipe roughness is 0.0002 and that the pump/motor efficiency is 65%. Estimate the pressure drop over the pipe line ignoring entry effects, also the power required.

The level of accuracy of the process information indicates that three term partitioning of the universes of discourse of the parameters as shown in Figure Ex. 2.6, Figure Ex. 2.7 and Figure Ex. 2.8 would be appropriate.

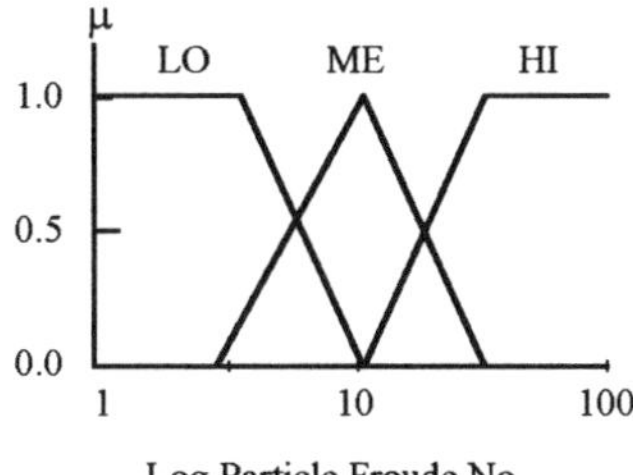

Figure Ex.2.6 Partitioning of the Particle Froude No.
LO = Low ME = Medium HI = High

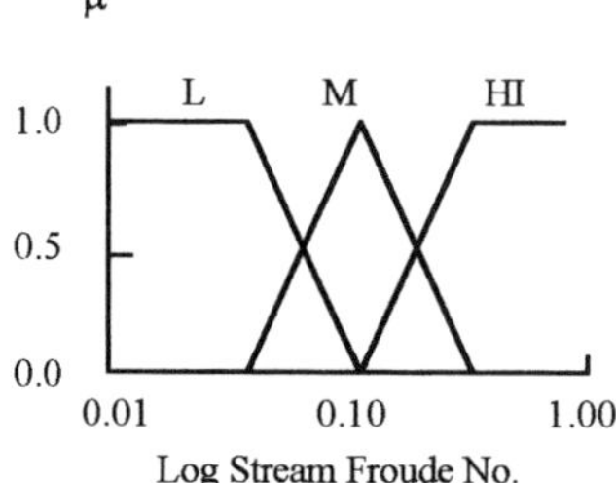

Figure Ex.2.7 Partitioning of the Stream Froude No.
LO = Low ME = Medium HI = High

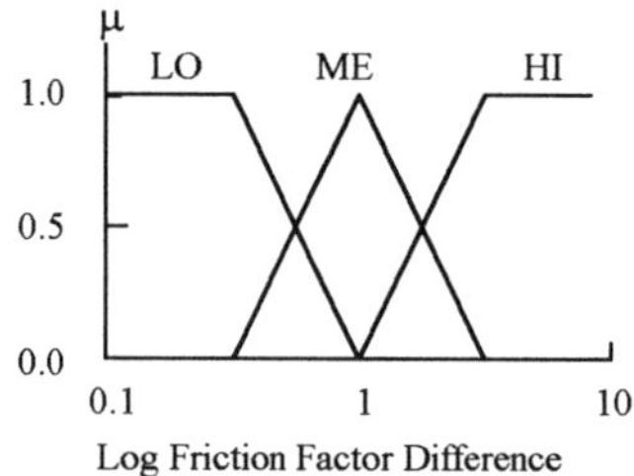

Figure Ex. 2.8 Partitioning of the Friction Factor Difference
LO = Low ME = Medium HI = High

Solution

It is first necessary to find the membership values of the particle and the stream Froude numbers from the above diagrams. From the given data

Suspension superficial velocity; v = Volumetric flow rate/Pipe area
$= 4*0.0139/\pi*0.07625^2 = 3.044$ m/s

Stream Froude No. $S = v^2/gD\rho'$
$= 3.044^2/9.81*0.07625*3 = 4.129$

Particle Froude No. $P = u^2/gd\rho'$
$= 0.108^2/9.81*0.002*3 = 0.198$

These values may be applied to Figure Ex. 2.6 and 2.7 to obtain the membership values of the linguistic sets

$S = 4.128$	$P = 0.198$
$\mu_{LO} = 0.77$	$\mu_{ME} = 0.35$
$\mu_{ME} = 0.23$	$\mu_{HI} = 0.65$

Combining these values with the rule-base given in Table 2.1 and applying the fuzzy proposition, results in the following conclusions

IF P	AND S	THEN F	MIN	CONCLUSION
LO	ME	HI	0.77,0.35	0.35 HI
LO	HI	HI	0.77,0.65	0.65 HI
ME	ME	ME	0.23,0.35	0.23 ME
ME	HI	HI	0.23,0.65	0.23 HI

The union of the above partial conclusions is shown below in Figure Ex. 2.9.

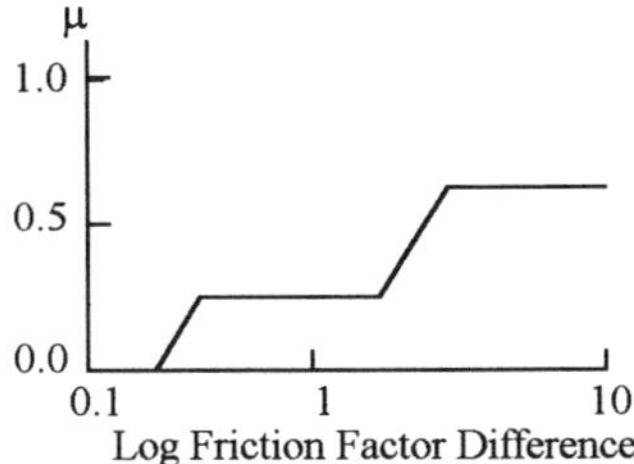

Figure Ex 2.9 Fuzzy Friction Factor Difference

The fuzzy friction factor may be defuzzified by the centroid method to give a value of 4.6.

The Reynolds Number for water alone with the same volumetric flow rate of 0.0139 m^3/s (assuming a viscosity of 0.001163 Pas) is

$$R' = 1000*3.05*0.07625/0.001163 = 200\ 000$$

From a standard friction factor-Reynolds number chart, assuming a pipe roughness of 0.0002, the friction factor is F' = 0.004125. From equation 2.13) the suspension friction factor is therefore

$$F = F'(1+\delta F) = 0.004125*5.6 = 0.0231$$

The pressure drop over the pipeline is given by

$$\delta p = 2\rho F v^2 l/D = 2*1179*0.0231*3.044^2*300/0.07625 = 199 \text{ kPa}$$

From which the power required is readily calculated to be 4.25 kW.

2.5 Mixing and dispersion

In the processing of materials there is frequently a need to introduce a mixing or dispersion process. Such a process is also known under a variety of other names such as; blending, stirring, beating, calendering or digesting, involving a wide range of substances from very low viscosity liquids which (if miscible) are very easy to disperse, to semi-solids and rubbery materials which are very difficult to mix and require high specific power inputs.

General deterministic solutions to problems involving power consumption and degree of dispersion or mixing, are difficult to find because of the variety of non-Newtonian behaviour encountered and the complexity of the flow patterns within the mixing vessel. In the chemical, pharmaceutical and foodstuffs industries mixing may also be accompanied by chemical reaction or heat transfer which further complicate the problem. The overall process performance evaluation then relies to a great extent on the formulation of dimensionless groups and empirical deterministic formulae.

A great variety of equipment geometries has evolved over an extended period of time to satisfy the needs of different industries. A typical example of a specialised type of equipment is the cement mixer of the construction industry.

It is also possible to classify mixing processes on different scales which represent different physical mixing mechanisms, and to a certain extent to relate these to the different geometries. This provides some rational explanation why different industries have developed different types of equipment. Various mixing scales and the associated mechanisms are shown below in Table 2.2

Table 2.2 Mixing Scales and Mechanisms

Mixing Scale	**Mechanism**
Molecular	Molecular diffusion
Micro	Small-scale turbulence
Mixer blade or ribbon scale	Vortices
Impeller scale	Secondary flow
Vessel scale	Primary flow

The relative size of the impeller, tank and material volume also play an important role in the mixing scales present and whether or not stagnation zones exist. The majority of industrial mixing processes are performed in cylindrical tanks, with or without baffles and the main geometric differences are in the form of the impeller and also its position in the tank. Usually, mixing is on a batch basis, although it is also sometimes arranged for the process streams to be continuous with either a static in-line-mixer with fixed baffles in a tube or using a stirred tank with continuous feed and product lines, plus a feedback loop for product enrichment.

Many different impeller designs are used in practice with single, twin or occasionally, multiple spindles. A selection of the main types, together with their usual operating characteristics in terms of their flow patterns, is given below in Table 2.3 for normal rotational speeds.

Table 2.3 Mixer Flow Patterns

Impeller Type		**Flow Pattern**		**Fluid Viscosity**
	Laminar	Vortex	Turbulent	
Turbine	0.1	0.3	0.6	Low
Propeller	0.2	0.4	0.4	Low
Anchor	0.2	0.6	0.2	Medium
Paddle	0.2	0.6	0.2	Medium
Helical Ribbon	0.8	0.2	0	High
Screw	0.8	0.2	0	High

In the above Table, the three columns under Flow Pattern show the approximate proportions of the various flow regimes. Also shown in the right hand column is the type of material that is usually associated with the given impeller type.

A complicating factor in the matching of an impeller geometry with fluid properties is that the properties may undergo a reversible or an irreversible change as the process proceeds. Thus a system may, for example, start as two-components with medium viscosities and during mixing it may evolve into a single-component material with high viscosity and perhaps also a yield stress. It may therefore be advisable to change the speed of the impeller during the mixing process, this could be achieved using a FL controller, (see Chapter 5).

Besides minimising capital plant costs it is also desirable to minimise the specific process energy, which affects the recurrent operating costs. Broad guidelines can be obtained by formulated by studying the literature, which mostly refers to laboratory and pilot-scale data, this can be merged with experience in a FL treatment to create a knowledge base.

The dimensionless power consumption (P) of the mixing process is customarily presented in the form of a power number-Reynolds number chart, which represents the relationship

$$P = \mathrm{f(R)} \qquad 2.19)$$

where P is the power number and R is the Reynolds number for a particular geometry. For the mixing process, these numbers are defined as follows

$$P = \mathrm{P}/s^3d^5\rho \qquad 2.20)$$

$$\mathrm{R} = \rho s d^2/\mu \qquad 2.21)$$

In equations 2.20) and 2.21), P is the impeller power input, s is the impeller rotational speed, d is the effective impeller diameter, ρ and μ are the average fluid density and viscosity respectively.

For a simple non-Newtonian power-law fluid the Reynolds number has an equivalent form as follows

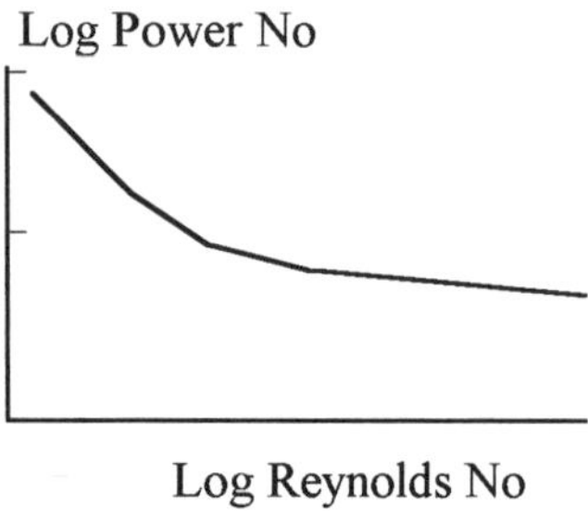

Figure 2.5 Typical Effect of the Impeller Reynolds Number on the Power Number

A typical power number chart is shown in Figure 2.5. The left-hand part of the curve represents the region where flow is entirely laminar. The highest Reynolds number region represents the region where energy dissipation is predominantly through turbulence. In the intermediate region, energy dissipation is due to secondary and vortex flow. The shape of the curve resembles that of the friction factor- Reynolds number chart for flow through packed columns. In the laminar region it is found that an empirical expression of the following form is approximately valid for a power-law type fluid

$$PR = A(n) \qquad 2.23)$$

where A(n) is a function of the flow index n and is constant for a given fluid system and impeller-tank geometry, (the fluid volume must also be a fixed proportion of the tank volume).

In practice it is difficult to make use of the published data because mixer geometries are not standardised and because of the range of fluid properties encountered. As an indication, some representative values of A(n) for n = 1 are given in Table 2.4.

Table 2.4 Indicative Values of A(1)

Impeller Type	**A(1)**
Helical screw with draught tube	270
Asymmetric helical screw, no tube	180
Anchor	175
Helical ribbon	120-300

It appears that for various fluids with several different impeller geometries, the critical power number for the departure from laminar flow conditions is between 5 and 10. Experience shows that for turbine and helical screw impellers the critical power number is about 10, whilst for helical ribbon impellers it is about 5. Various categories of design could obviously be characterised by FL sets.

The helical ribbon mixer has a strong axial pumping action and therefore it is a reasonable conjecture that the degree of mixing is proportional to the number of impeller revolutions. The work done by the impeller is divided between creating fresh interfacial area between the material components and friction. On dimensional grounds the specific work done, (w), may be expressed as P/sd^3. Now from equations 2.20) and 2.22)

$$PR = P/ks^{n+1}d^3 = (P/s^2d^3)/ks^{n-1} \quad 2.24)$$

Combining equations 2.22) and 2.23), the unit work may be defined as

$$w/s = P/s^2d^3 = A(n)ks^{n-1} \quad 2.25)$$

The interpretation of equation 2.25) is that with a pseudo-plastic fluid ($n<1$) there is a tendency for the unit work to decrease as the speed of rotation increases, provided that n and k remain constant. It will be recalled that A(n) is dependent upon the impeller- tank geometry. As the impeller speed increases it will be found that beyond a certain point vortices will be formed, followed by turbulence, the effect will be to reverse the trend of diminishing unit work. The general pattern will be as illustrated in Figure 2.6.

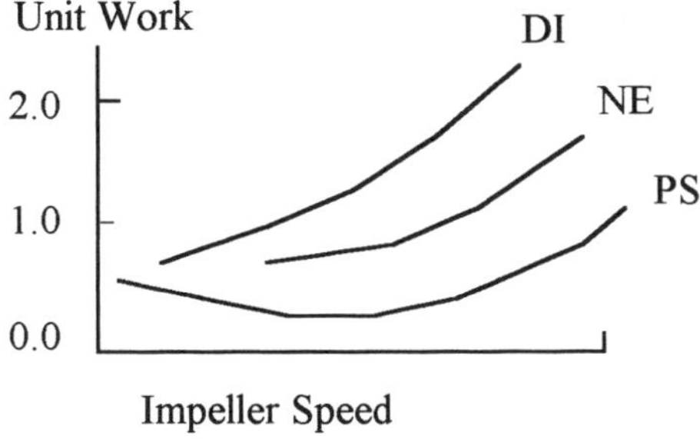

Figure 2.6 The General Trend of the Effect of Impeller Speed on Unit Work for Various Types of Fluids. DI = Dilatant. NE = Newtonian.
PS = Pseudoplastic.

The indication given from this simple analysis is that for a pseudoplastic fluid the minimum unit work condition is obtained by operating just beyond the laminar flow conditions. For a Newtonian fluid it appears as if there is no minimum condition at finite impeller speeds. In practice, these indications would need to be merged with experience of similar systems and weighted with a plausibility factor to create a robust rule-base.

Within each type of mixer there will be found in practice many combinations of sizes and whilst Table 2.4 shows sample values of A(1) for a number of different types of mixer (for Newtonian fluids), Table 2.5 shows sample values for various combinations of dimensions for the helical ribbon type of mixer (again for Newtonian fluids).

Table 2.5 A(1) Values for Helical Ribbon Mixers

Mixer	d/D	p/d	h/d	c/d	w/d	No.Ribbons	A(1)
1	0.898	0.517	1.01	0.0575	0.135	1	230
2	0.912	0.495	0.942	0.0485	0.0971	1	207
3	0.912	1.00	0.996	0.0485	0.0971	2	250
4	0.902	1.00	1.01	0.0539	0.0981	1	130
50	0.910	1.00	1.00	0.050	1.00	2	246

In the above Table; d = impeller outer diameter, D = tank diameter, h = ribbon axial length, c = (D-d)/2 and w = ribbon width.

It may be noted from the above Table that there is a 63% variation about the mean value of A(1) and therefore for a given helical ribbon mixer which may not exactly match any of the above five combinations some similarity metric is needed to make use of the data. One such similarity metric is defined by the following min-max expression

$$m = \Sigma^{q}_{i=1} x_i / \Sigma^{q}_{i=1} y_i \qquad 2.26)$$

where m is the metric value, x_i is the trial value and y_i is the corresponding tabulated parameter value. The following example illustrates the method of comparison of a sample mixer with the specifications given in Table 2.5.

Example 2.4

A commercially produced helical ribbon mixer has the following geometry

d/D	p/d	h/d	c/d	w/d	
0.9	0.7	1.0	0.0051	0.11	Single Ribbon.

It is required to blend two liquid feedstocks in the volume ratio of 1:2, the feedstocks have viscosities of 8.14 and 1.48 Pa s respectively. Estimate the specific work for the blending process.

If the tank charge is 0.1 m^3 and it takes 35 impeller revolutions to produce a product of the required quality, estimate the total work required.

Solution

Comparing the given geometry with the data in Table 2.4 by using the metric expression, equation 2.25) the most similar tabulated mixer may be found. Only mixers 1,2 and 4 have single ribbons and therefore the others need not be considered.

$$m_1 = (0.898+0.517+1.0+0.051+0.11)/(0.9+0.7+1.01+0.0575+0.135)$$
$$= 0.9192$$

$$m_2 = (0.9+0.495+0.942+0.0485+0.0971)/(0.912+0.7+1.0+0.051+0.11)$$
$$= 0.8953$$

$$m_4 = (0.9+0.7+1.0+0.051+0.0981)/(0.902+1.0+1.01+0.0539+0.11)$$
$$= 0.8938$$

The highest similarity metric value is obtained for the first mixer which has an A(1) value of 230. The weighted mean of the feedstock viscosities is

$$\mu_m = (8.14+2*1.48)/3 = 3.7 \text{ Pa s}$$

The specific work is $w = 230\mu_m = 230*3.7 = 851$ J/m^3/revolution

The total work required is W = w*Tank charge*No.revolutions
= 851*0.1*35 = 2979 J

In the turbulent region it is found that the power number,*P*, tends towards a constant value, say *B*, which is independent of the Reynolds number as illustrated in Figure 2.5, thus

$$P/\rho s^3 d^5 = B \qquad 2.27)$$

The specific work in the turbulent region is therefore

$$w = B\rho s d^2 \qquad 2.28)$$

Experience shows that *B* is a function of impeller geometry and has a value of about 4.5 for a turbine impeller and about 0.35 for paddle and anchor impellers.

Following the treatment of flow through packed columns and porous media, the expressions for laminar and turbulent flow, equations 2.25) and 2.28) above, are simply added together to yield

$$w = A(n)ks^{n-1}+B\rho s d^2 \qquad 2.29)$$

The minimum value, which is indicated in Figure 2.6) for pseudoplastic fluids, is found by differentiating equation 2.29)

$$dw/ds = (n-1)A(n)ks^{n-2}+B\rho s d^2 = 0 \qquad 2.30)$$

Hence $$s_{min} = [(1-n)A(n)k/B\rho d^2]^{1/(2-n)} \qquad 2.31)$$

Which gives the impeller speed for the minimum specific work condition.

Example 2.5

A non-Newtonian aqueous suspension of flocculated china clay is found from viscometric laboratory data to be represented in laminar shearing motion by shear-thinning behaviour with a flow index of 0.62, a flow coefficient of 5.5 Pa s, and it also a specific has a specific gravity of 1.2. It is to be agitated using a paddle type impeller, for which it is estimated using a min-max metric with available data that A(n) = 60 and *B* = 0.8. The impeller swept diameter is 0.5 m. Estimate the impeller speed for minimum energy consumption per unit volume of product.

Solution

The minimum specific energy consumption is given directly by equation 2.30). Using the given data with this expression

$$s_{min} = (0.38*60*5.5/0.8*0.5^2*1200)^{1/1.4}$$
$$= 0.625 \text{ rev/sec or } 37.5 \text{ rev/min.}$$

This result has an implied level of accuracy which exceeds that of the supporting data. It is found in practice that there is significant scatter in the available experimental data, resulting in a degree of uncertainty in both A(n) and *B*, and therefore in the value of the specific work.

As with many industrial process, there is a significant level of uncertainty in the predicted performance of mixers, the sources of some contributory factors are outlined in the above discussion. This means that a FL treatment is more representative of the knowledge environment than a deterministic one. For a particular type of agitator the parameter A(n) may be related to the degree of departure from non-Newtonian behaviour by suitable partitioning of the universes of discourse and a rule-base. Experience indicates that for an anchor agitated vessel this might be expressed as shown in Tables 2.6, 2.7 and 2.8 below.

Table 2.6 Flow Index Difference. (1-N)

	(1-N)					
Set.	0	0.1	0.2	0.3	0.4	0.5
NN	0	1	0	0	0	0
SN	0	0	1	0	0	0
MO	0	0	0	1	0	0
LN	0	0	0	0	1	0

where NN = Near Newtonian, SN = Small non-Newtonian, MO = Moderately non-Newtonian and LN = Large non-Newtonian.

Table 2.7 Mixer Parameter A(N).

	A(N)					
Set	35	57	86	115	150	170
SM	0	1	0	0	0	0
ME	0	0	1	0	0	0
LM	0	0	0	1	0	0
LA	0	0	0	0	1	0

where SM = Small, ME = Medium, LM = Large Medium, LA = Large

Table 2.8 Rule-Base

(1-N)	NN	SN	MO	LN
A(N)	LA	LM	ME	SM

Example 2.6

Consider the same dispersion problem as in *Example 2.5*, but with the (1-N), A(N) and rule-base as shown in Tables 2.6, 2.7 and 2.8 respectively. Also consider the fuzzy set form of B given by

$$B = [0.5//0.672+1.0//0.800+0.5//0.928]$$

Find the impeller speed for minimum specific energy consumption for this case.

Solution

With the same value of the flow index as before, then (1-n) = 0.38. Using this value to interpolate in Table 2.6 gives the following membership values

$$\mu_{MO} = 0.26$$
$$\mu_{LN} = 0.74$$

The Rule-Base in Table 2.8 provides the corresponding fuzzy set of A(N) which may be entered into a FL proposition form to give

IF (1-N)	THEN A(N)	MEMBERSHIP VALUE	CONCLUSION
MO	ME	0.26	0.26 ME
LN	SM	0.74	0.74 SM

Finding the union of the partial conclusions and defuzzifying, weighting by the membership values, the resulting A(n) value is

$$A(0.62) = 0.26*86+0.74*57 = 64.54$$

Considering equation 2.30), all the terms in this expression are deterministic except B. Evaluating these terms

$$(1\text{-}n)A(0.62)k/\rho d^2 = 0.38*64.54*5.5/1200*0.5^2$$
$$= 0.45$$

Using this result together with the given fuzzy value to find the impeller speed for minimum specific work, using the extension principle

$$s_{min} = [0.45/B]^{1/(2-0.62)}$$
$$= [0.5//0.670+1.0//0.563+0.5//0.485]^{0.725}$$
$$= [0.5//0.748+1.0//0.659+0.5//0.591] \text{ rev/sec}$$

Defuzzifying this result using weighted values gives

$$s_{min} = 0.66 \text{ rev/sec. or about 40 revs/min.}$$

The A(n) relationship in the above is actually non-linear and is may be noted how easily the FL method absorbs this without modelling. The power of approximate reasoning is shown in this case and where there is appreciable scatter in the physical data and hence uncertainty.

CHAPTER 3

THERMAL PROCESSES

Engineering thermodynamics has its roots in classical thermodynamics; this subject is concerned with a system, the boundary of which is a continuous surface defined by the observer, and which is in thermodynamic equilibrium with its surroundings. Thermodynamic equilibrium means; thermal, chemical and mechanical (also more generally, gravitational, magnetic and electrical) equilibrium.

The thermodynamic state of the system is defined by its thermodynamic properties such as: temperature, pressure, specific volume, entropy and enthalpy. The thermodynamic properties of a substance are related by empirical equations of state , which are valid over limited ranges and verified by experiment, or by tables of experimentally determined values. Thermodynamic properties are only defined in a state of equilibrium, although they are also customarily used in discussions about practical cases which are invariably concerned with processes which may be far from equilibrium. Instrument recordings of properties may then only be approximately in accord with the predictions of theory.

Systems are recognised as being either open or closed. If there is no mass transfer across the system boundary, the system is said to be closed, otherwise it is open.

Ideal processes are discussed in classical thermodynamics, but they occur in systems passing through equilibrium states and are therefore reversible, devoid of any frictional effects, temperature gradients or any other non-ideal conditions that would cause irreversible entropy changes due to the entropy of the universe (system plus surroundings) increasing. These processes are unattainable in reality and serve as limits to actual processes, which are always irreversible.

From the above discussion it will be readily appreciated that there are a number of reasons why uncertainty is present in calculations relating to real thermodynamic processes and why, even under the best possible physical conditions, the observations may only correspond approximately with theoretical calculations. In general, thermodynamic properties are then more properly represented by fuzzy sets.

The main concern here is in the beneficial effect of improved definition in solutions of thermal process problems through the use of FL methods. The customary treatment is through the application of the laws of thermodynamics and empirical physical models. In one type of alternative FL treatment the empirical models are discarded in favour of a FL rule-base. In another, the empirical deterministic models become fuzzy-deterministic. In either case solutions are found in the form of fuzzy sets. Intelligent design, control and management of thermal processes requires a realistic knowledge base.

The customary treatment will normally yield a solution in terms of a single numerical value which is “about right”, but not exact. In the FL treatment a broader meaning is given to the solution and it contains more information It is therefore of higher quality than the deteministic solution. The fuzzy set solution may be defuzzified to yield a single value, but the background information remains.

A more complete discussion of thermodynamic aspects would also include reference to irreversible thermodynamics of quasi-equilibrium states and the Onsager relations, with reciprocal coefficients. Also to statistical thermodynamics and stochastic processes. These are excluded in the present text, they do not fall within the usual compass of engineering thermodynamics.

3.1 Non-flow processes

Classical thermodynamics theory is based upon the study of closed (non-flow) systems. When such a system is subject to sequential processes so that it returns to its original state, then a cycle is defined. In the ideal case of the theory, the cycle is reversible so that the system and its surroundings (the universe) may be restored to their original state by reversing the processes.

There are four known laws of thermodynamics: The zeroth law defines temperature. The first law defines the internal energy of a system (which for the ideal gas is a function only of temperature). The second law defines entropy, which is a thermodynamic property of the system. The third law defines the zero of entropy, which in engineering thermodynamics is not of significance because only entropy differences are required.

If the process undergone by a system is reversible then the entropy change of the system must be equal and opposite to the entropy change of its surroundings. In other words, the nett change of entropy of the universe is zero. In engineering texts "reversible" frequently means internal reversibility of the system, without reference to the surroundings. and in this case the process will, in general, yield a nett change of entropy. Entropy changes are found either by using tables of thermodynamic properties, or by the assumption of deterministic empirical equations for the internal energy and the equation of state (usually for an ideal gas in engineering thermodynamics). These empirical equations may be found to approximately model real systems over limited ranges of the properties, provided no phase changes occur.

Besides the above properties and specific volume, a number of other derived thermodynamic properties are known, these important quantities, which are derived in terms of the primary properties already mentioned, are: enthalpy, Gibb's free energy and the Helmholtz free energy. They are discussed in many of the standard texts on classical thermodynamics.

It may be noted that the entire basis of classical thermodynamics rests on CL. It is when the pure theory is applied to real processes that, for the reasons already discussed, approximations are needed and uncertainty become apparent. Then factors are introduced to reconcile observations with theory.

Batch processes in chemical engineering, although not specifically treated here, are also in the category of non-flow processes.

The following example illustrates the use of the extension principle (see Appendix) for processing physical data in a fuzzy-deterministic formulation of the frequently used polytropic expansion model. It is an alternative to using several local rule-bases patched together using logical union to cover the range of variables of interest.

Example 3.1

A closed cylinder contains a homogeneous charge of gas which is mainly air, but also contains significant amounts of gaseous contaminants, the amounts varying from one charge to another. The total mass of gas in the cylinder is 0.5 kg and it is initially at a pressure of 15 bar and a temperature of $250^{o}C$. The gas expands slowly to a pressure of 1.5 bar. The index of expansion is not known exactly, because of the variation in the amounts of contaminants, but it is considered that the index may be represented by a fuzzy set

$$N = [0//0.9n+0.5//0.95n+1.0//n+0.5//1.025n+0//1.05n]$$

where $n = 1.25$.

Assuming that the gas constant may be represented by

$$R = [0//0.273+0.5//0.280+1.0//0.287+0.5//0.294+0//0.301] \quad kJ/kgK$$

Also that $C_v = [0//0.682+0.5//0.700+1.0//0.718+0.5//0.736+0//0.754]$ kJ/kgK

Estimate the work done by the gas during expansion, also the heat transfer and the entropy change.

Solution

Assuming that the expansion process may be modelled by a reversible fuzzy polytropic process. Let suffix 1 denote initial conditions and suffix 2 denote final conditions, then

$$T_2 = T_1(P_2/P_1)^{(N-1)/N}$$

This expression may be evaluated using the extension principle.

Now $$N-1 = [0//(0.9n-1)+0.5//(0.95n-1)+1.0//(n-1)+0.5//(1.025n-1)+ 0//(1.05-1)]$$

Taking the principal set

$$(N-1)/N = [0//(0.9n-1)/0.9n+0.5//(0.95n-1)/0.95n+1.0//(n-1)/n+ 0.5//(1.025n-1)/1.025n+0//(1.05n-1)/1.05n]$$

If $n = 1.25$ then

$$(N-1)/N = [0//0.777+0.5//0.695+1.0//0.631+0.5//0.603+0//0.578]$$

Now, $T_2 = (273+250)(15/1.5)^{(N-1)/N} = 523(10)^{(N-1)/N}$

Therefore by using the above fuzzy index

$$T_2 = [0//302+0.5//315+1.0//330+0.5//363+0//406]$$

The expression for the work done may be shown to be

$$W = mR(T_1-T_2)/(N-1)$$

where m is the mass of gas.

Now, $T_1-T_2 = [1.0//523]-[0//406+0.5//363+1.0//330+0.5//315+0//302]$
$= [0//117+0.5//160+1.0//193+0.5//208+0//221]$ K

and $(T_1-T_2)/(N-1) = [0//936+0.5//851+1.0//772+0.5//740+0//706]$ K

Also, the work done per unit mass is

$$W/m = [0//255+0.5//238+1.0//222+0.5//218+0//213]\text{kJ/kg}$$

Given that, m = 0.5 kg, then the work done is

$$W = [0//107+0.5//109+1.0//111+0.5//119+0//178] \text{ kJ}$$

The defuzzified value of the work done by the gas during expansion is 112.5 kJ.

The first law of thermodynamics may be expressed as

$$Q\text{-}W = U_2\text{-}U_1$$

Also, assuming that the internal energy is only a function of temperature, the heat transferred is therefore

$$Q = W+mC_v(T_2-T_1)$$

$$= [0//107+0.5//109+1.0//111+0.5//119+0//178]+[1.0//0.5]* [0//0.682+0.5//0.700+1.0//0.718+0.5//0.736+0//0.754]* [0//117+0.5//160+1.0//193+0.5//208+0//221] \text{ kJ}$$

The above expression yields thirty terms. The principal fuzzy set from the Cartesian products gives five elements comprising an envelope which when illustrated as a graph contains the remaining twenty five components. The principal fuzzy set is

$$Q = [0//23+0.5//32+1.0//41+0.5//63+0//138] \text{ kJ}$$

The heat transferred to the gas contains values in the range from 23 kJ to 138 Kj to some degree about the value of 41 kJ. The defuzzified value is 44.25 kJ.

The envelope of Q represented, as a continuous distribution, is shown in Figure Ex 3.1. (See the Appendix for a discussion of principal sets).

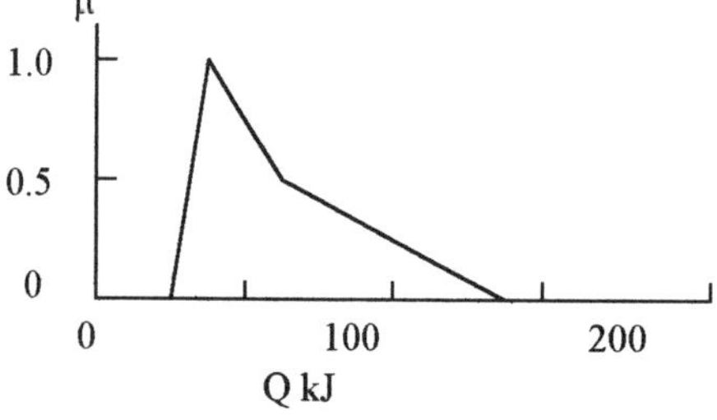

Figure Ex 3.1 The Heat Transfer (Q) Fuzzy Set

It may be shown that for a polytropic reversible process the change of entropy is given by

$$S_2\text{-}S_1 = (C_v\text{+}R/(1\text{-}N))\ln(T_2/T_1)$$

Now, $\ln(T_2/T_1) = [0//\ln(406/T_1)+0.5//\ln(363/T_1)+1.0//\ln(330/T_1)+0.5//\ln(315/T_1)+0//\ln(302/T_1)]$
$= [0//-0.253+0.5//-0.365+1.0//-0.460+0.5//-0.507+0//-0.549]$

$R/(1\text{-}N) = [0//-2.18+0.5//-1.49+1.0//-1.15+0.5//-1.04+0//0.962]$ kJ/kg

and $C_v+R/(1\text{-}N) = [0//-1.50+0.5//-0.790+1.0//-0.432+0.5//-0.286+0//-0.208]$ kJ/kgK

Hence, $s_2\text{-}s_1 = [0//0.114+0.5//0.145+1.0//0.199+0.5//0.288+0//0.380]$ kJ/kgK

or $S_2\text{-}S_1 = [0//0.057+0.5//0.073+1.0//0.100+0.5//0.144+0//0.190]$ kJ/K

The change of entropy of the gas during expansion is in the range of 0.057 kJ/K to 0.190 kJ/K, about the value of 0.100 kJ/K. The defuzzified value is 0.133 kJ/K.

Note that in the above, the principal sets of Cartesian products are again taken. Solutions using a rule-based approach will look different on a chart, but should have comparable defuzzified values and distribution tendencies.

3.2 Otto, Diesel and compound cycles

Originally, piston engines were exclusively driven by steam and were thus external combustion engines. But this practice is now reduced to special applications. At the present time (1999) piston engines are almost entirely of the internal combustion type, in which fuel is burnt under pressure inside the cylinder to liberate heat, inducing a still higher pressure and thus causing the piston to deliver mechanical work to the surroundings.

There are a few other types of external combustion (including solar-powered) piston engines, the best known are those based upon the Stirling cycle and to a lesser extent, the Ericsson cycle. These too are currently limited to special applications. Both the two latter types are special because the ideal thermal efficiencies are equal to the theoretical limit set by the Carnot cycle for given upper and lower operating temperatures. No real engines of whatever type display thermal efficiencies that are equal, or even close to, the Carnot efficiency; they all exhibit significant irreversibilities.

The Carnot cycle has a low work ratio, for this and other reasons it is not used as a basis for real engines, such as the ubiquitous spark-ignition (SI) and compression-ignition (CI) types. There are several theoretical reversible cycles that are more similar to these practical engines, they are known as the air-standard power cycles and are used as benchmarks for the thermal efficiency of practical cycles. They assume that the working fluid is air, that the heat is added to and removed from the air by external sources of heat and cold and that the working fluid is expanded and compressed isentropically. The thermal efficiencies given in most standard engineering thermodynamics texts for three common air-standard cycles are listed below:

Otto cycle. Heat is supplied and removed at constant volume. The thermal efficiency is

$$\eta = 1\text{-}1/r_v^{0.4}$$

Diesel cycle. Heat is supplied at constant pressure and removed at constant volume. The thermal efficiency is

$$\eta = 1\text{-}(r_c^{1.4}\text{-}1)*(1.4(r_c\text{-}1))*1/r_v^{0.4}$$

Compound cycle. Heat is partly supplied at constant volume and partly at constant pressure. Heat is removed, as in the previous case, at constant volume, the thermal efficiency is

$$\eta = 1\text{-}(r_p r_c^{1.4}\text{-}1)/((r_p\text{-}1)+1.4r_p(r_c\text{-}1))*1/r_v^{0.4}$$

where $r_v = v_1/v_2$ (Figure 3.1a)) ; $r_p = p_3/p_2$ (Figure 3.1c))

and $r_c = v_4/v_3$ (Figure 3.1b))

The Otto, Diesel and compound cycles are shown in Figure 3.1.

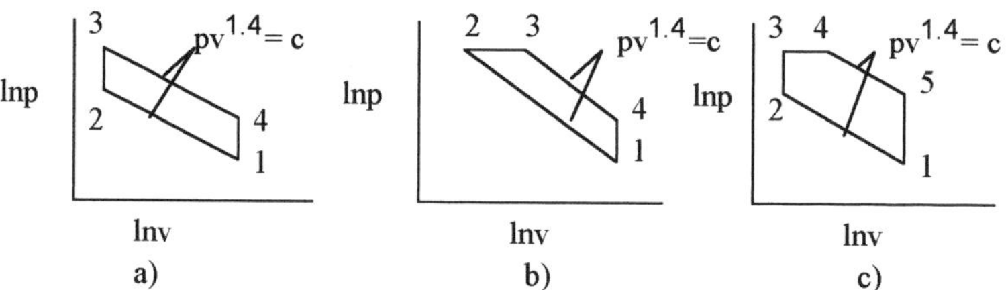

Figure 3.1 a) Otto cycle. b) Diesel cycle. c) Compound cycle.

All the above air standard cycles have thermal efficiencies that are lower than the Carnot cycle working between the same temperature limits.

The Otto cycle is considered to be similar to the SI engine cycle and the high-speed CI engine cycles, whilst the Diesel cycle or the combined cycle is considered to be similar to the low and medium-speed CI engine cycle. No account is taken of the variation of air properties with temperature (and of lesser importance, the pressure) in the theoretical treatments.

Although there are similarities to real engine cycles, there are also important differences. These are aspirated and have air/fuel and combustion products as working fluids, the mixtures are not entirely homogeneous and in practice only surface measurements of temperature, pressure and the piston movement are observable. There is also turbulence, swirling within the cylinder and heat transfer to and from the solid boundaries. Another factor is effect of the finite opening and closing times of the inlet and exhaust valves. All these factors provide significant departures from the ideal reversible cycles mentioned above, and limits to their predictive powers for real engine cycles.

Actual combustion temperatures may reach 2500K to 2800K, at which temperatures dissociation of gas molecules will occur (with some recombination as the temperature falls). This will also have an effect on gas properties.

The effect on the three gas cycles is that the index of expansion and compression may be different to the usually quoted value of 1.4 (the value for air at 288K and at low pressure). Also the specific volume ratio of the gas may be different to the value obtained by measuring the piston travel, the mixture in a real cylinder may also be inhomogeneous.

The following example, based on a fuzzified Otto cycle, is a further use of the extension principle and allows for uncertainty in the Otto cycle parameters in a non-ideal case.

Example 3.2

In a fuzzy Otto cycle the index of expansion and compression of the gas is considered to be best represented by a fuzzy set "about 1.21" for a range of engine speeds

$$N = [0//1.1+1.0//1.21+0.5//1.3+0//1.4]$$

Also the specific volume compression ratio is given by another fuzzy set "about 8"

$$R_V = [0//6.4+0.5//7.2+1.0//8.0+0//9.0]$$

Find the corresponding fuzzy thermal efficiency and compare it with the ideal Otto cycle air standard efficiency.

Solution

From the given data

$$N\text{-}1 = [0//0.1+1.0//0.21+0.5//0.3+0//0.4]$$

Using the extension principle

$$R_V^{N-1} = [0//1.204+0.5//1.484+1.0//1.516+0.5//1.866+0//2.408]$$

The fuzzy Otto cycle efficiency is given by

$$= [1.0//1]-[R_V^{1-N}]$$
$$= [1.0//1]- [0//0.831+0.5//0.674+1.0//0.660+0.5//0.536+0//0.415]$$
$$= [0//0.169+0.5//0.326+1.0//0.340+0.5//0.464+0//0.585]$$

The continuous representation of the fuzzy efficiency set is illustrated in Figure Ex 3.2. It contains all values of efficiency between 16.9% and 58.5% to some degree, clustered around the value of 34%. The defuzzified value, taking into account the spread, is 36.75%. The air standard efficiency based upon a volume ratio of 8 and the usual index value of 1.4 is 56.47%.

The volumetric efficiency, and hence R_V measured by the specific gas volume, is dependent upon engine speed in a real engine.

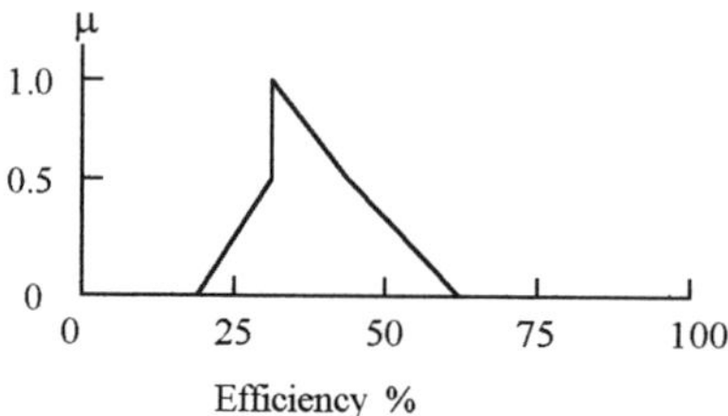

Figure Ex. 3.2 Otto Efficiency Fuzzy Set

The work ratio, defined as the ratio of the nett cycle work to the gross expansion work is also important in evaluating the ideal cycle; for the Carnot cycle it is invariably low.

3.3 Engine characteristics

Real engine performance studies are based upon the correlation of empirical data, guided in general terms by classical thermodynamics theory. The value of mathematical models is small because of the physical complexities, it is difficult to capture the range and vagueness of actual data. The linguistic basis of fuzzy logic can however summarise and portray experience without the restrictions of classical logic and deterministic mathematical models.

The main SI and CI performance metrics are:
i) Indicated and brake mean effective pressure (imep and bmep respectively)
ii) Specific fuel consumption. (sfc)
iii) Power/engine volume ratio
iv) Mechanical efficiency
v) Exhaust emissions
vi) Piston speed

Some of these metrics are combined in the form of performance maps which summarise results for a range of engine series, or even engines of different models. It is clear that these maps should really have fuzzy boundaries.

The above metrics are mainly influenced by:
i) Engine geometry (and to a lesser extent, engine size)
ii) Fuel characteristics. The calorific value of petroleum fuels varies only slightly with different grades and types. More important is the octane number (SI engines), or the cetane number (CI engines).
iii) Ignition timing (SI engines) or injection timing (CI engines)
iv) Valve timing
v) 2 or 4-stroke principle
vi) Air/fuel ratio

vii) Ambient conditions (Atmospheric temperature, pressure and humidity)
viii) Engine wear

Any engine will usually require an initial preservice running-in period under light load. Engine performance usually refers to the subsequent period, before inevitable wear and consequent performance deterioration is apparent. Ambient conditions can affect the performance by up to about 6% for each factor. To establish comparability of performance data, various standards exist such as; ISO 3046-3, 1989, BS 5514, Part 3, 1990, DIN 6270 and SAE J1349, which prescribe reference test conditions. But there is still need for further rationalisation.

Engine geometry, particularly the compression ratio and the combustion chamber design, has an important effect on performance. Swirling and turbulence can promote combustion efficiency, but excessive turbulence leads to rough engine running, which is mechanically undesirable. This is affected by combustion chamber design.

For a given engine series, the bmep, bsfc and piston speed are closely related. There are no models available for these relations, in some cases performance maps are available, but whether these are available or not a rule-base may be constructed based on expert opinion. In general terms, they might have a rule-base as illustrated in Table 3.1 and Table 3.2. (SI and CI engines respectively). The L,M and H entries in the two tables would imply ranges that would be lower for the CI engine than for the SI engine.

Table 3.1 SI Rule-Base for BSFC

	EP			
PS	L	LM	HM	H
L	H	M	L	M
M	H	H	M	M
H	H	H	H	

Table 3.2 CI Rule-Base for BSFC

	EP			
PS	L	LM	HM	H
L	H	M	L	M
M	H	M	M	H
H	H	H	H	H

The above abbreviations have the following meanings:

EP = bmep
PS = Piston speed
BSFC = Brake sfc

L = Low
LM = Low medium
M = Medium
HM = High medium
H = High

Example 3.3

A new modification to an existing engine series is under consideration. The engine series performance data is summarised in the fuzzy diagrams in Figure Ex.3.3 given below. Assuming the same rule-base given in Table 3.1. Estimate the bsfc for a piston speed of 7.5 m/s and a bmep of 5 bar.

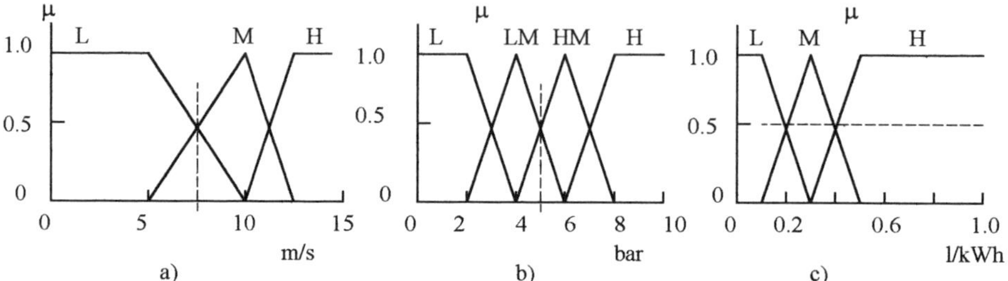

Figure Ex. 3.3 Engine Performance Fuzzy Diagrams
a) Piston Speed b) Brake Mean Effective Pressure c) Brake Specific Fuel Consumption

Solution

In this case the problem is model-free, the solution is by inference by fuzzy propositions using logical union of partial conclusions. The antecedents, or inputs, are the piston speed and the bmep. The table of propositions is given below. The linguistic fuzzy sets are found from Figure Ex 3.3 using the given inputs.

IF PS	AND EP	THEN BSFC		MIN	INFERENCE
L	LM	M	0.5,0.5	0.5 M	
L	HM	L	0.5,0.5	0.5 L	
M	LM	H	0.5,0.5	0.5 H	
M	HM	M	0.5,0.5	0.5 M	

The inference is shown on Figure Ex 3.3c). The defuzzified value is 0.5 l/kWh.

3.4 Thermo-flow processes

Consideration is now given to steady flow processes in which the local time derivatives of all variables is zero. Following a convected element of mass through a process, the thermodynamic definitions, relations and laws are assumed to apply to the element, as for the non-flow process discussed earlier. For steady flow, the thermodynamic properties at a fixed point in the flow region are constant. The first law of thermodynamics referred to a fixed co-ordinate system needs to be generalised; the energy content of a unit mass of material now includes not only the internal energy as discussed in Section 3.1, but also kinetic energy, potential energy and a pressure-volume term, which may be called pressure energy. This latter quantity is combined with the internal energy to define the quantity called enthalpy.Using specific quantities

$$h = u+pv \qquad 3.4)$$

The steady flow energy equation, quoted here without proof is

$$q_{12}-w_{12} = h_1-h_2 \ + \text{(kinetic energy and potential energy terms)} \qquad 3.5)$$

This is the equivalent of the first law of thermodynamics for an open system, where suffices 1 and 2 refer to sequential process positions of the material element. q_{12} is the heat transfer and w_{12} is the work transfer per unit mass. Neglecting the heat transfer, work transfer and potential energy terms, then the gain in kinetic energy per unit mass is

$$[C_2^2-C_1^2]/2 = H_1-H_2 = C_p[T_1-T_2] \qquad 3.6)$$

Friction effects are accommodated in the steady flow energy equation 3.5) and it therefore applies both to reversible and irreversible processes. In the solution of problems using equation 3.5), it is normally assumed that at transverse points across the flow lines the thermodynamic and kinematic properties are constant. This is only an approximation because element velocities will vary to some extent, being zero at the surface of a stationary solid boundary. The assumption of constant thermodynamic properties cannot be exactly true.

Because of the non-ideal factors mentioned above, and the concommital element of vagueness, the customary way of solving flow process problems is to firstly assume an ideal reversible process, then to factor the solution by a process efficiency, which is an assumed correction factor, based upon experience. A FL treatment of the same problem dispenses with the latter step because the fuzzy set description of the data already accommodates the uncertainty.

In the case of flow processes, some physical modelling is possible and this produces better correspondence with reality with the use of fuzzy models.

Example 3.4

A gas receiver is maintained at a pressure of 3.5 bar ± 5% and the gas has a temperature of 160^oC ± 10% over a period of time. The gas velocity in the receiver is negligible. The gas expands through a small orifice to a pressure of 1.0 bar. Assume that the orifice acts as a nozzle and that the index of expansion is 1.35 ± 5%. The gas specific heat at constant pressure is Cp = 1.006 kJ/kgK.

Estimate the gas exit velocity.

Solution

Let suffices 1 and 2 refer to the receiver and the nozzle exit plane respectively. Then from equation 3.6)

$$[C_2^2 - C_1^2]/2 = H_1 - H_2 = Cp[T_1 - T_2]$$

The left-hand side of equation 3.6) represents the change of kinetic energy per unit mass. Now $C_1 =$ 0 also assuming a polytropic expansion, then the exit gas temperature is given by

$$T_2 = T_1(P_2/P_1)^{(N-1)/N}$$

From the given data, the index of expansion is

$$N = [0//1.28 + 0.5//1.32 + 1.0//1.35 + 0.5//1.38 + 0//1.42]$$

and the initial pressure is

$$P_1 = [0//3.32 + 0.5//3.41 + 1.0//3.5 + 0.5//3.59 + 0//3.68] \text{ bar}$$

Using the extension principle, the principal fuzzy set of (1-N)/N is

$$(1-N)/N = [0//-0.358 + 0.5//-0.422 + 1.0//-0.473 + 0.5//-0.524 + 0//-0.596]$$

It follows that the pressure ratio raised to the power (N-1)/N is given by the principal fuzzy set; on the right-hand side of the following equation

$$(P_2/P_1)^{(N-1)/N} = [0//0.460 + 0.5//0.512 + 1.0//0.552 + 0.5//0.600 + 0//0.651]$$

The initial temperature is given by

$$T_1 = [0//417 + 0.5//425 + 1.0//433 + 0.5//441 + 0//449] \text{ K}$$

The exit temperature principal fuzzy set is

$$T_2 = [0//192 + 0.5//218 + 1.0//243 + 0.5//265 + 0//292] \text{ K}$$

The exit velocity is

$$C_2 = (2C_p(T_1 - T_2))^{1/2}$$

The principal fuzzy set of the velocity velocity is therefore

$$C_2 = [0//501 + 0.5//567 + 1.0//618 + 0.5//670 + 0//719] \text{ m/s}$$

The fuzzy set for the exit velocity represented as a continuous distribution is shown in Figure Ex 3.4. It may be noted that although the initial data fuzzy sets are symmetrical, the exit velocity is not. The exit velocity contains all values between 501 m/s and 719 m/s to some degree.

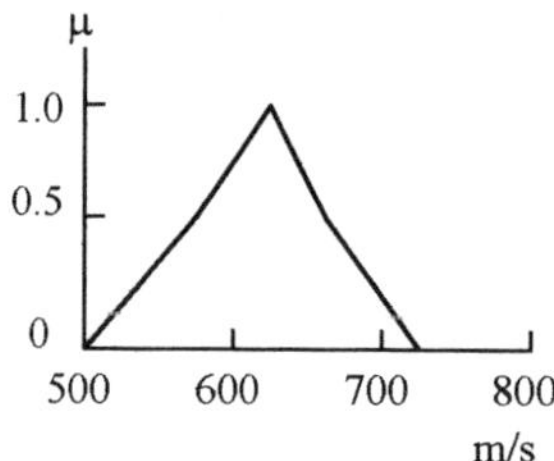

Figure Ex 3.4 Fuzzy Exit Velocity (C_2).

3.5 Joule cycle

This is another example of an air-standard cycle.

The Joule (or Brayton) cycle comprises the following sequence of processes:

i) Isentropic compression to the upper pressure.
ii) Heating at the constant upper pressure.
iii) Isentropic expansion to the lower pressure.
iv) Cooling at the constant lower pressure.

The air standard efficiency of the ideal Joule cycle is

$$\eta = 1 - (1/r_p)^{(\gamma-1)/\gamma} \quad 3.6)$$

The corresponding work ratio (r_w) is

$$r_w = 1 - (T_1/T_3)r_p^{(\gamma-1)/\gamma} \quad 3.7)$$

where r_p is the ratio of upper to lower cycle pressure.

The Joule cycle is the reference for closed and open cycle gas turbines. In contrast to piston engines, they aspirate relatively large volumes of air and the upper cycle temperature is significantly lower. It has a higher internal circulation of power and therefore the work ratio is generally lower. There is also considerably more turbulence in the working fluid and hence the compression and expansion are strongly irreversible adiabatic processes, rather than isentropic. In the representation of this cycle it is usually assumed that the end states of compression and expansion may be joined by a suitable reversible process.

Example 3.5

Consideration is given to a gas turbine design to work under variable conditions. It is to be modelled on the Joule cycle. The pressure ratio is assumed to be about 6.5 and to have the following profile

$$R_p = [0/5.85+0.5/6.18+1.0/6.5+0.5/6.83+0/7.15]$$

The lower and upper temperatures are about 293K and about 950K respectively

$$T_1 = [0/273+0.5/283+1.0/293+0.5/308+0/333]$$
$$T_3 = [0/855+0.5/903+1.0/950+0.5/998+0/1045]$$

The index of expansion is assumed to be 1.4.

Estimate the thermal efficiency and the work ratio.

Solution

In FL form the thermal efficiency is

$$\eta = [1-(1/R_p)^{(\gamma-1)/\gamma}]$$

where $(\gamma-1)/\gamma = 0.286$

From the given R_p set, the extension principle yields

$$(1/R_p)^{0.286} = [0//0.603 + 0.5//0.594 + 1.0//0.585 + 0.5//0.577 + 0//0.570]$$

Hence the efficiency profile is

$$\eta = [0//0.397 + 0.5//0.406 + 1.0//0.415 + 0.5//0.423 + 0//0.430]$$

Thus the efficiency ranges from 39.7% to 43.0% about the value of 41.5%.

Considering the principal fuzzy sets

$$T_1/T_3 = [0//0.261 + 0.5//0.283 + 1.0//0.308 + 0.5//0.341 + 0//0.389]$$

and $$R_p^{0.286} = [0//1.658 + 0.5//1.684 + 1.0//1.709 + 0.5//1.733 + 0//1.754]$$

also $$(T_1/T_3)R_p^{0.286} = [0//0.433 + 0.5//0.477 + 1.0//0.526 + 0.5//0.591 + 0//0.682]$$

The fuzzy set form for the work ratio is

$$R_w = [1- (T_1/T_3)R_p^{0.286}]$$

The work ratio profile is therefore

$$R_w = [0//0.318 + 0.5//0.409 +1.0//0.474 + 0.5//0.523 + 0//0.567]$$

Thus the work ratio range is 0.318 to 0.567 about the value of 0.474. The work ratio profile is illustrated in Figure Ex 3.5. It may be noted that the R_w range is much greater than that for η.

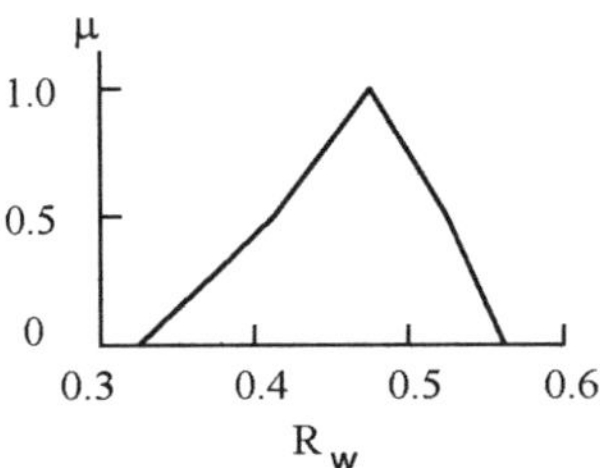

Figure Ex. 3.5 The R_w Profile

3.6 Rankine cycle

This is the usual model for vapour power cycles, of which the steam power plant is the most common example. Typical basic features of the cycle are illustrated on the temperature-entropy diagram in Figure 3.2.

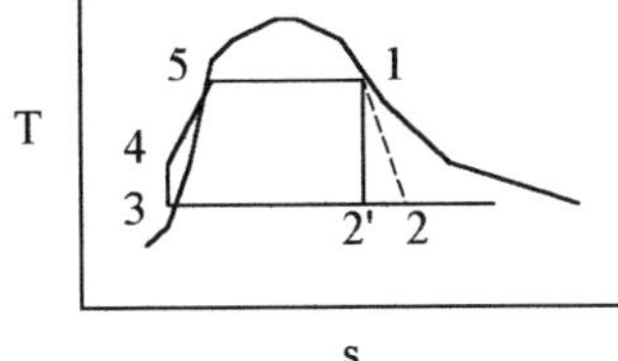

Figure 3.2 The Rankine Cycle on a Temperature-Entropy Diagram

The ideal cycle comprises an isentropic expansion of the working fluid from a dry saturated vapour at state point 1, to a wet vapour at the lower pressure, stage 2. The vapour is condensed at constant pressure to a saturated liquid, state 3. A feed pump then delivers the liquid to a boiler at the upper pressure, state point 4. At this point the liquid is heated at constant pressure to the saturated liquid, state point 5, and further heated at constant pressure to the dry saturated vapour state point 1.

The feed pump work (3-4) is usually negligible compared to the expansion work, (1-2) (or 1-2'). The work ratio for the Rankine cycle is normally close to unity. For this reason the feed pump work is often ignored in calculations.

There are many possible modifications to the basic Rankine cycle such as superheating the working fluid beyond state point 1, reheating the working fluid at some intermediate turbine pressure and also by bleeding vapour from the turbine during the expansion process. These are all discussed in standard engineering thermodynamics texts.

In a steam plant the criteria of performance are evaluated by various metrics such as: thermal efficiency, specific steam consumption (ssc) (kg/kWh) which enables the relative size of different steam plant to be compared. Also the efficiency ratio, which is the ratio of the actual plant efficiency to that of the equivalent Rankine cycle efficiency for steam plant and the isentropic efficiency for gas turbines are frequently encountered. At the design stage, the latter quantity is assumed from previous experience with similar plant under similar operating conditions.

The isentropic efficiency allows for departure from the ideal expansion end point 2' to the actual supposed operating point 2, due to friction heating of the working fluid. In practice this may vary under different working conditions.

The expansion end points 1 and 2 may be related by a polytropic relation of the form

$$p_1 v_1^n = p_2 v_2^n \quad 3.8)$$

Isentropic efficiency is implied, but in practice there is uncertainty in the value of n, as in the isentropic efficiency.

Example 3.6

A steam turbine plant is planned to drive an electricity generating set. The electrical load demand is uncertain at any future point in time, but it is estimated that the profile may be expressed as

$$L = [0//0.6F + 0.5//0.8F + 1.0//F + 1.0//1.2F + 0//1.4F]$$

where F depends upon the season of the year.

The steam boiler supply is dry saturated at a pressure of 40 bar. The condenser pressure is 0.030 bar. The state points at the beginning and end of expansion are related by a polytropic expansion, and to allow for uncertainty in the end condition the polytropic index is expressed as a fuzzy set

$$N = [0//1.085 + 1.0//1.088 + 0.5//1.090 + 0//1.093]$$

Neglecting boiler feed pump work, find the cycle efficiency and the specific steam consumption. Also find the steam consumption profile if the seasonal load factor is 5MW.

Solution

From tables of thermodynamic properties the following specific steam properties at the turbine inlet may be found:

Specific enthalpy h_1 = 2801 kJ/kg. Specific volume v_1 = 0.04977 m^3/kg. Specific entropy s_1 = 6.070 kJ/kgK.

At the turbine outlet:

Saturated liquid enthalpy h_{f2} = 100.6 kJ/kg. Latent enthalpy h_{fg2} = 2444kJ/kg. Dry saturated specific volume v_{g2} = 45.92 m^3/kg. Saturated liquid entropy s_{f2}= 0.352 kJ/kgK. Latent entropy s_{fg2} = 8.226 kJ/kgK.

Let suffix 2 of the primed specific properties represent the turbine outlet conditions following an isentropic expansion.

Hence $\quad s_1 = s'_2 = s_{12}+x'_2 s_{fg2}$

The dryness fraction after isentropic expansion is therefore

$$x'_2 = (s_1-s_{f2})/s_{fg2} = (6.070-0.352)/8.226 = 0.695$$

The ideal specific volume is $\quad v'_2 = x'_2 v_{g2}$
The actual specific volume is $\quad v_2 = x_2 v_{g2}$

Hence the actual dryness fraction is $x_2 = x'_2(v_2/v'_2)$
where v_2 is found from the polytropic relation

$$p_1 v_1^N = p_2 v_2^N$$

From the given data $x_2 = 0.695v_2/31.91 = 0.02178v_2$

and $\quad v_2 = v_1(p_1/p_2)^{1/N} = 0.04977(40/0.03)^{1/N} = 0.04977*13333.3^{1/N}$

The above expressions may be evaluated in tabular form, as given below. The extension principle has been used in the calculations.

μ	**n**	**1/n**	**$13333.3^{1/N}$**	**v_2 m^3/kg**	**x_2**	**h_2 kJ/kg**
0	1.085	0.9217	759	37.78	0.823	2111
0.5	1.088	0.9191	745	37.08	0.808	2075
1.0	1.090	0.9174	736	36.63	0.798	2050
0	1.093	0.9149	723	35.98	0.784	2016

In the above table, h_2 is found from; $h_2 = h_{f2} + x_2 h_{fg2} = 100.6 + 2444x_2$ kJ/kg

In the following table, the thermal efficiency (η) and the specific steam consumption (ssc) are evaluated using.

Enthalpy differences, $h_1 - h_2 = 2801 - h_2$ kJ/kg

$$h_1 - h_3 = 2801 - 100.6 = 2700 \text{ kJ/kg}$$

Thermal efficiency $\qquad \eta = (h_1 - h_2)/(h_1 - h_3) = 0.0003703(2801 - h_2)$ kg/kWh

Steam consumption $\qquad ssc = 3600/(h_1 - h_2)$ kg/kWh

μ	**h_1-h_2**	η	**ssc**
0	690	0.2556	5.217
0.5	726	0.2689	4.959
1.0	751	0.2781	4.794
0	785	0.2907	4.586

From the above tabulation, the efficiency and specific steam consumption profiles are

$$\eta = [0//0.2556 + 0.5//0.2689 + 1.0//0.2781 + 0//0.2907]$$

$$ssc = [0//4.586 + 1.0//4.794 + 0.5//4.959 + 0//5.217] \text{ kg/kWh}$$

With a load factor, F, of 5 MW the load profile is

$$L = [0//3 + 0.5//4 + 1.0//5 + 1.0//6 + 0//7] \text{ MW}$$

The steam consumption is given by

$$Q = [1.0//10^3]L/(H_1 - H_2) \text{ kg/s}$$

This expression is calculated in tabular form below.

	L MW				
H_1- H_2 kg/s	0//3	0.5//4	1.0//5	1.0//6	0//7
0//690	4.348	5.797	7.246	8.696	**10.140**
0.5//726	4.132	5.560	6.887	**8.264**	9.642
1.0//751	3.995	**5.326**	**6.658**	7.989	9.321
0//785	**3.822**	5.096	6.369	7.643	8.917

The principal fuzzy set, shown in bold type in the above table, is determined from the maximum and minimum values at each level of the membership function, (0,0.5 and 1.0).

The steam consumption profile is therefore

$$Q = [0//3.822 + 0.5//5.326 + 1.0//6.658 + 1.0//7.989 + 0.5//8.264 + 0//10.140] \text{ kg/s}$$

3.7 Heat exchangers

Fluid - Fluid heat transfer is a widely encountered industrial process, and it is commonly accomplished by passing the fluids through a heat exchanger. The three main types of heat exchanger are:

i) *Recuperators.* The two fluids exchanging heat are separated by a wall. Usually one fluid flows within a metal tube and the other fluid passes over the external surface.
ii) *Regenerators.* The two fluids are passed alternately through a matrix, which one fluid heats and the other cools, thereby effecting heat transfer. If the matrix is rotated in the fluid streams it is possible to create continuous heat transfer.
iii) *Evaporative.* In this type the two fluids are in direct contact, so that both heat and mass transfer occur simultaneously.

There are also other important types of heat transfer such as fluid-solid and radiant heat transfer. In the following discussion only recuperators will be considered, they are probably the most common. Most analyses of this type of heat exchanger consider a model comprising two coaxial tubes with one of the fluids passing through the inner tube and the other fluid passing axially through the annular space between the two tubes. The bulk mean temperature profiles are shown in Figures 3.3a) and b) for the cocurrent and countercurrent cases respectively.

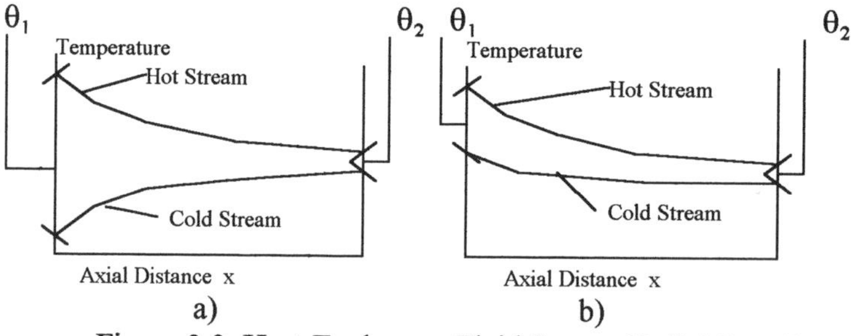

Figure 3.3 Heat Exchanger Fluid Stream Bulk Mean Temperature Profiles
a) Concurrent Flow
b) Countercurrent Flow

For the case where one of the fluids is a condensing vapour, the corresponding temperature profile is horizontal on the diagrams; there is no distinction between the cocurrent and countercurrent cases.

The basic heat exchanger formula for the heat transfer rate, q, is

$$q = ua\theta' \tag{3.9}$$

where u is an overall heat transfer coefficient, a is the contact area between the fluid streams and θ' is the effective mean temperature difference between the fluid streams.

θ' is often defined as the logarithmic mean temperature difference (LMTD)

$$\theta' = (\theta_1 - \theta_2)/\ln(\theta_1/\theta_2) \qquad 3.10)$$

There are several important simplifying assumptions in the derivation of θ': the heat transfer coefficient, u, is constant throughout the heat transfer process, (which generally it is not), also that there are no heat losses, (which there are), and that the radial heat conduction rate is much greater than the axial conduction rate, which it often is. θ_1 and θ_2 are the bulk mean temperature differences, which are themselves idealisations, there will be a temperature profile across each fluid stream which will depend upon the nature of the flow regime.

Simple heat exchangers of the coaxial tube type are not often found in practise, multi-pass, cross-flow and mixed flow types are much more common. Extended or finned surfaces are also frequently used to enhance the effective contact area, a, in equation 3.9). under these circumstances an analytical derivation of the effective mean temperature difference, θ', and also a precise determination of the effective contact area, a, is not possible. The method usually adopted is to calculate a LMTD based on counterflow, using the same entry and exit temperature differences, and to apply to this an empirical correction factor, y, so that the estimated effective temperature difference, θ'' is given by

$$\theta'' = y\theta' \qquad 3.11)$$

Graphs of y for different heat exchanger configurations will be found in process heat transfer texts. They are given in terms of hot and cold side temperature ratios and fluid stream heat capacity ratios. Industrial fluid heat transfer is nearly always designed to occur under turbulent flow conditions due to the enhancing effect of turbulent eddies, though this does incur the penalty of higher fluid pumping power requirements. In turbulent flow the resistance to heat transfer resides mainly in the fluid film adjacent to a solid boundary because the heat transfer through the film is by conduction. There will also be some resistance offered by the metal tube wall, but this is kept to a minimum.

Neglecting the curvature of the tube wall the overall heat-transfer factor u' is given by

$$1/u' = 1/u_1 + t/k + 1/u_2 + R \qquad 3.12)$$

where u_1 and u_2 are the film heat-transfer coefficients, t is the tube wall thickness, k is the thermal conductivity of the tube metal and R is a fouling factor which is zero for a clean tube, but which generally increases in value depending on the deposition rate of solids on the tube wall from the fluid streams.

Defining a heat exchanger hot-stream efficiency as the ratio of the fall in temperature of the hot stream to the temperature difference between the hot stream inlet and cold stream inlet. For a particular heat exchanger geometry, the efficiency will be given in functional form by

$$e = f(w_1,w_2,a,u) \qquad 3.13)$$

where w_1 and w_2 are the heat capacities (w = mc) of streams 1 and 2 respectively. a and u are as previously defined.

By dimensional analysis, the following functional relationship is obtained

$$e = f(w_1/w_2,au/w_2) \qquad 3.14)$$

or $$E = f(W,A) \qquad \text{where } W = w_1/w_2 \text{ and } A = au/w_2 \qquad 3.15)$$

A similar result is obtained for the cold stream efficiency.

In equation 3.15) the symbols may now be interpreted as fuzzy sets. If the rule-base is known for E ,W and A and the universes of discourse are partitioned in a suitable way, then the efficiency may be determined. This enables one of the temperatures of the heat exchanger to be determined provided that the other three are known.

It has been noted that there are reasons why deterministic modelling of practical heat exchangers is difficult. In the following example the dimensionless groups are related by a rule-base to provide a model-free treatment.

Example 3.7

In a chemical processing plant a stock liquor at 305K is to be heated using water at 380K, which is discharged from another process. The heat exchanger has an effective heat transfer area of 19.5m and the transfer mode is mainly counter-current. The specific heats for the liquor and water are 2.3kJ/kgK and 4.2kJ/kgK respectively. The overall heat transfer factor is estimated to be 450W/m K. The flow rate of liquor is 3.0kg/s and that of water is 2.0kg/s.

Find the exit temperature of both streams.

The following rule-base may be used for the efficiency, E.

	A		
W	S	M	L
S	ME	HI	HI
M	ME	ME	HI
L	LO	ME	ME

The partitioned universes of discourse are shown in Figure Ex. 3.7.

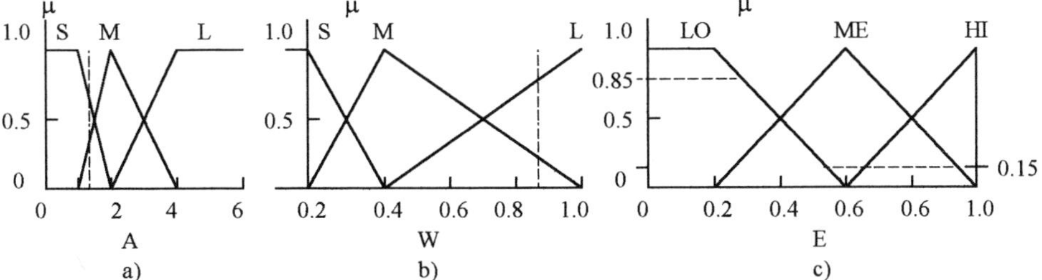

Figure Ex. 3.7 Partitioning of the Universes of Discourse
a) Fluid Heat Capacity Ratio, A b) Dimensionless Heat Transfer Rate, W c) Efficiency, E

Solution

From the given data $w_1 = 3.0*2.3 = 6.9$ kW/K

and $w_2 = 2.0*4.2 = 8.4$ kW/K

Therefore $W = w_1/w_2 = 0.8214$

also, $A = 19.5*450/8.4*10^3 = 1.045$ m

From the fuzzy diagrams, Figure Ex 3.7, the membership values are:
For A = 1.045; membership values, 0.95S and 0.05M.
For W = 0.8214; membership values, 0.15M and 0.85L.

Using the given rule-base and the above membership values the following table of propositions and conclusions may be compiled

IF A	AND W	THEN E	MIN	INFERENCES
S	L	LO	0.95,0.85	0.85 LO
S	M	ME	0.95,0.15	0.15 ME
M	L	ME	0.05,0.85	0.05 ME
M	M	ME	0.05,0.15	0.05 ME

The fuzzy set for the efficiency is shown on Figure Ex 3.7. Defuzzifying the diagram using the centroid method, the value of the resulting efficiency is e = 0.335. The efficiency has high membership in the 0 to 25% range and tapers off to a low membership beyond 55%.

Let the hot and cold stream temperatures be defined by t' and t'' respectively, then the hot stream efficiency is given by

$$e = (t_1' - t_2')/(t_1' - t_1'')$$

or,

$$0.335 = (380 - t_2')/(380 - 305)$$

Hence the hot stream (water) outlet temperature is 354.8K.

Assuming no heat loss

$$w_1(t_1' - t_2') = w_2(t_2'' - t_1'')$$

Hence the cold stream (liquor) outlet temperature is

$$t'' = (380 - 354.8)*6.9/8.4 + 305 = 325.7K$$

Note that in Figure Ex. 3.7 the computer grid sets a limit on the accuracy of portrayal of the input intersections with the fuzzy sets in diagrams a) and b).

3.8 Heat transfer coefficient

It may be recalled that in turbulent flow the main resistance to heat transfer is due to the fluid film which exists at any solid boundary, (except perhaps under rarefied gas conditions). For heat transfer in a tube under laminar flow conditions an analytical solution is possible and therefore a FL treatment is not relevant. But if turbulent conditions prevail, film heat transfer coefficients are experimentally determined and the results are empirically presented in terms of dimensionless groups. It has been found from much experimental data that for a range of geometries, the heat transfer data may be correlated in dimensionless group form as follows.

$$N_u = KR_e^m P_r^n \qquad 3.16)$$

where N_u= hd/k (Nusselt number); R_e= vd/ν (Reynolds number); P_r= cν/k (Prandtl number).

In the above, h is the film heat transfer coefficient, d is the characteristic dimension (usually the tube diameter), k is the thermal conductivity of the fluid, ρ is the fluid density, ν is the fluid viscosity, c is the heat capacity and v is the mean fluid velocity. In equation 3.16) K, m and n are constants that depend upon the flow geometry and the type of fluid (i.e.,gas, Newtonian fluid or non-Newtonian Fluid). The reference conditions for evaluating the constants should be ascertained when using published values.

In equation 3.16) the fluid properties are often taken as those at the mean film temperature which itself is not easy to determine in industrial equipment; they are average values in some sense. Liquid viscosity may be quite sensitive to temperature in some cases, moreover in non-Newtonian fluids there is no unique viscosity because it is shear rate depended (and may also be dependent upon the flow history). These factors make the Reynolds number and the Prandtl number difficult to define with certainty. It is possible to define an alternative relationship to equation 3.16) which is not explicit in the viscosity and therefore avoids some of the associated problems. This is achieved by defining the following alternative dimensionless groups

$P_e = R_eP_r$ = cρvd/k (Peclet number); $S_t = N_u/R_eP_r$ = h/ρvc = td/ 4θ'l (Stanton number); $F_f = JR_e$ = (dp'/4l)/(1/2ρv²). Where t is the fluid temperature change, p' is the pressure drop and l is the effective fluid path length. For non-circular ducts, d is assigned a value given by

$$d = 4*\text{flow cross-sectional area/wetted perimeter} \qquad 3.17)$$

The friction factor-Reynolds number relationship above is empirical, determined from experimental data. With the above definitions, equation 3.16) may be recast in the following form

$$S_t = K'F_f^{m'}P_e^{n'} \qquad 3.18)$$

or,

$$S_t = K''R_e \; P_r \qquad 3.19)$$

Typical values available for the constants in equations 3.16), 3.18) and 3.19) are given below in Table 3.3.

Table 3.3 Typical Values of Constants in Equations 3.16), 3.18) and 3.19)

	Newtonian Liq.		**Gas**		**External**
	Tube	**Annulus**	**Tube**	**Annulus**	**Transverse Flow**
K	0.023	0.023	0.020	0.018	0.34
m	0.8	0.8	0.8	0.8	1.6
n	0.4	0.37	0	0	0
K'	3.96 $*10^{-4}$	1.95 $*10^{-4}$	0.214	0.137	28.2
m'	-1.6	-1.88	0.8	0.8	1.6
n'	-0.6	-0.67	0	0	0
K''	0.023	0.023	0.028	0.029	0.48
m''	-0.2	-0.2	-0.2	-0.2	-0.4
n''	-0.6	0.67	0	0	0

There are also a few results available for non-Newtonian fluids, but these are generally less certain than for Newtonian fluids. Values of the constants given in Table 3.3 are usually obtained with clean metal surfaces under limited ranges of physical variables for heating of the internal flow. There is still some uncertainty in these results. For industrial equipment a FL approach, which is model free and able to incorporate published data and corporate experience, is more credible, and moreover avoids the need for assuming an exchanger efficiency. It may be noted in Table 3.3, that for a gas n, n' and n'' are all zero and therefore simplified forms of equations 3.16), 3.18) and 3.19) are possible. This means that a simplified form of FL proposition of the form; IF A THEN B is also possible.

Experience indicates that for gas flow a rule-base of the following form is suitable

F_f	SM	MS	MB	BI
S_t	SM	MS	MB	BI

where SM = small; MS = small medium; MB = big medium and BI = big.

This represents the FL relationships between the linguistic variables, incorporating the general trend of the system behaviour. Partitioning of the universes of discourse translates the linguistic variables into fuzzy sets. A similar rule-base to the above would be found for the R_e-N_u and the R_e-S_t combinations. A two-dimensional array would be appropriate for liquid systems.

Example 3.8

It is planned to preheat an air supply of 1.9 kg/s using a two-pass shell and tube heat exchanger utilising waste process steam at a rate of 0.0504 kg/s. The steam temperature is 370K and its latent heat is 1700 kJ/kg. The air blower delivery pressure for the given air flow rate should be 460 Pa and the air is to flow through the bank of heat exchanger tubes which are made from 25 mm diameter tube stock. Assume a mean air density of 1.11 kg/m and a specific heat at constant pressure of 1.005 kJ/kgK.

Estimate the effective air film heat transfer coefficient. Check the number of heat exchanger tubes and their length.

For the given heat exchanger geometry, the FL rule-base is

F_f	SM	MS	MB	BI
S_t	SM	MS	MB	BI

The corresponding partitioning of F_f and S_t is shown in Figure Ex 3.7

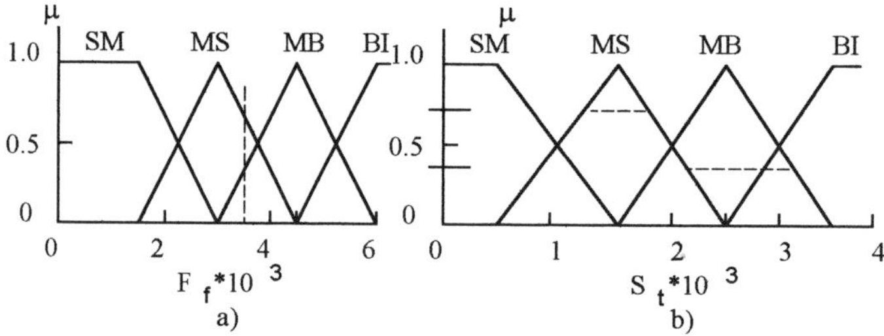

Figure Ex 3.7 Partitioning of the Universes of Discourse.
a) Friction factor F_f. b) Stanton number S_t.

Solution

The steam heat supply is given by $q = mh = 0.0504*1700 = 95.2$ kW

Air temperature rise $t = q/ac = 95.2/1.9*1.005 = 44.87$ K

Temperature differences $\theta_1 = 370\text{-}288 = 82$ K
$\theta_2 = 370\text{-}333 = 37$ K

From equation 3.10) $\theta' = (82\text{-}37)\ln(82/37) = 56.38$ K

Assuming a trial value of the friction factor, $F_f = 3.5*10^{-3}$

From Figure Ex 3.8a) and the rule-base, the inferences are

IF F_f	THEN S_t	MIN	INFERENCE
MS	MS	0.65	MS 0.65
MB	MB	0.35	MB 0.35

The fuzzy inference is shown in Figure Ex 3.8b). This may be defuzzified by the centroid method yielding a Stanton number, $S_t = 2.28*10^{-3}$.

Now, as noted previously $F_f = 0.5(d/l)p'/\rho v^2$

and $S_t = 0.25t(d/l)/\theta' = 0.25*44.87(d/l)/56.38$

or $S_t = 0.1990(d/l)$

Hence, $F_f/S_t = 2.26p'/v^2 = 3.5/2.28$

which yields $p'/v^2 = 0.6780$

Since p' is given as 460 Pa, v is easily found to be 26.05 m/s.

Now also $S_t = h/\rho vc$ (as before)

Hence, the film coefficient is $h = St\rho vc$

Therefore, $h = 2.28*10^{-3}*1.11*26.05*1.005*10^3$
$= 66.26$ W/mK

This appears to be a reasonable value for the film coefficient. To check whether it is consistent with an acceptable heat exchanger design, the corresponding number and effective length of tubes is calculated as follows.

The air mass flow rate is $a = \rho nxv$

where x is the cross-sectional area of one tube $= 4.91*10^{-4}$ m^2

Hence, $n = 1.9/1.11*4.91*10^{-4}*26.05 = 134$ tubes.

As found above, $S_t = 0.1990d/l$

Hence, $l = 0.1990*25*10^{-3}/2.28*10^{-3} = 2.18$ m

Both the number of tubes and the effective tube length would be suitable for a two-pass heat exchanger with a well supported tube bank. With condensing steam on one side of the exchanger, the heat transfer factor would be only slightly affected by the film coefficient.

For liquid-liquid heat transfer a two-dimensional rule-base would be required for the proposition.

3.9 Solar collectors

In recent times significant attention has been given to harvesting solar energy using technological devices, both as a means of reducing demands on fossil fuels in industrialised regions and as a means of supplying additional thermal and electrical energy in some of the less industrialised regions that are remote from the electrical grid. Solar energy use can reduce the demand for both fossil fuels and for fuel wood, though its diffuse nature and variability means that its range of uses is limited, especially in non-tropical regions.

To harvest solar energy, technological devices of many different designs have been created, these generally fall into one of two categories:

i) Photothermal, in which the radiation (sometimes focussed) is converted into heat.
ii) Photovoltiac, in which the radiation is converted directly into electrical energy.

Nature has evolved a very large number of interesting biological solar collectors in the form of plants and trees. Natural selection has ensured that these are well suited to their environment conditions, including the weather pattern. They differ from technological devices in that they are integrators of solar energy over periods ranging from months to tens or hundreds of years, whereas the technological devices normally store and discharge energy on a daily basis. The biological solar systems are not further considered here.

A technological solar energy system usually comprises a collector surface which converts the incident solar radiation into either thermal or electrical energy. This is then transferred to a storage unit for aggregation and later use. The most common examples of these are water heaters and battery chargers. The basic system is shown in Figure 3.4.

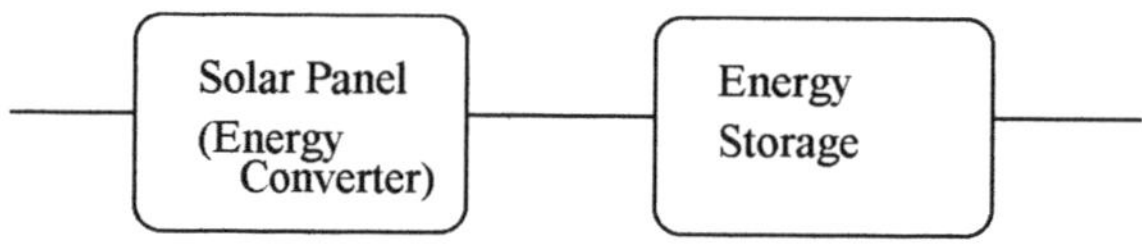

Figure 3.4 Basic Solar Energy System

The performance of these systems depends upon the design features, the installation features (such as orientation) and the weather pattern. Because of the latter factor the system performance varies on a daily as well as on a seasonal basis and there is an element of uncertainty, or vagueness, that depends upon the location and the accuracy of the weather forecasts. Generally, the longer term that the weather forecast is the more vague it becomes beyond the gross seasonal variations.

For a given system design, location and installation the main factors affecting daily performance are the time of the year and the daily cloud cover. Provided that a standard demand pattern is accepted, then the performance of a solar energy system may be cast in fuzzy logic form. In the case of a solar water heater the demand pattern may, for example, be the maximum volume of water drawn per 24 hours at a temperature of 45°C. In the case of a photovoltiac device it may be the amount of electrical energy drawn at 12 volts in a 24 hour period. Either of these represent a standardised output. The current British Standard (1986) uses a draw-off temperature of 55°C.

Let CC represent the cloud cover, SE the season and OP the output, then the fuzzy logic proposition for a given system is

IF CC AND SE THEN OP

By partitioning the cloud cover, season and output universes of discourse, the fuzzy logic representation may be as shown in Figure 3.5.

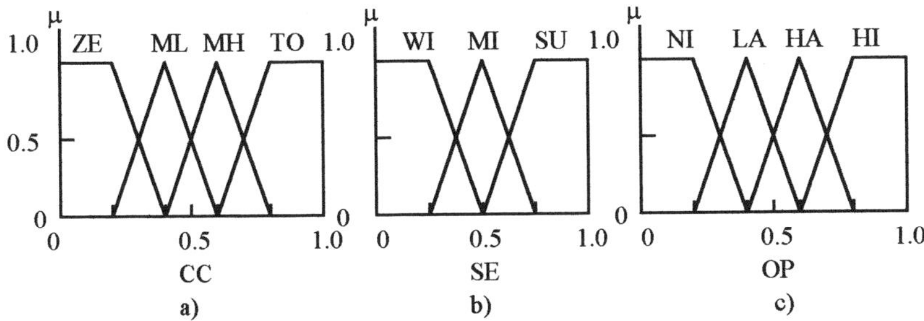

Figure 3.5. Partitioning of the Normalised Universes of Discourse.
a) Cloud Cover,CC b) Season,SE c) Output,OP

In Figure 3.5 the linguistic terms have the following meanings:

a).ZE Zero	b) WI Winter	c) NI Nil
ML Medium low	MI Mid-season	LA Low average
MH Medium high	SU Summer	HA High average
TO Total		HI High

Thus 1.0 TO represents total dense cloud cover. 1.0 SU represents mid-summer. 1.0 MI represents mid-season, which is the mid-point between mid-winter and mid-summer. 1.0 HI represents the highest standardised output under the most favourable conditions for that particular location. The relationship between the cloud cover, season and standardised output is embedded in the rule-base. Table 3.4 gives an example of a rule-base.

Table 3.4 Rule-Base for a Solar System.

	CC			
SE	TO	MH	ML	ZE
WI	NI	NI	NI	LA
MI	NI	NI	LA	LA
SU	NI	LA	HA	HI

For a tropical region, the sun passes directly overhead twice a year, therefore the rule-base would need to be amended.

Example 3.9

A solar water heater can deliver 150 litres of water per day at 45°C under ideal clear sky conditions in mid-summer. The installation is such that it is only able to deliver 80% of the ideal amount of energy. The weather forecast is that there will be sunny spells and cloudy periods on 01 September, which is early autumn. Water supply to the solar heater is at 15°C. Assume that the fuzzy diagrams in Figure 3.5 and the rule-base in Table 3.4 apply, also that the average specific heat of the water is 4.19 kJ/kgK.

Estimate the likely hot water output on 01 September if it is decanted at 45°C.

Solution

The standard delivery is

$$u = mc_pT = 150*4.19*10^3*(45-15)$$
$$= 18.86 \text{ MJ/day}$$

The weather forecast indicates a mixture of sun and cloud, which means that without more detailed information one would tend to classify conditions as medium cloud cover. A satellite picture would assist in refining the forecast and enable a choice to be made between medium low and medium high cloud cover. A compromise would be to select the cross-over value between these two conditions, which on the fuzzy diagram, Figure Ex3.9a) would mean an average of 50% cloud cover for the active period of the solar collector; for a fixed installation under normal conditions this is about two hours before solar noon to about three hours after solar noon.

Taking 22 December as zero on the normalised universe of discourse of the season, Figure Ex3.9b), and 22 June as unity, then 01 September is day 253. On the normalised scale this equates to 2*112/365 = 0.6137. Inserting this value in Figure Ex3.9b) gives the following membership values: 0.55 MI and 0.45 SU.

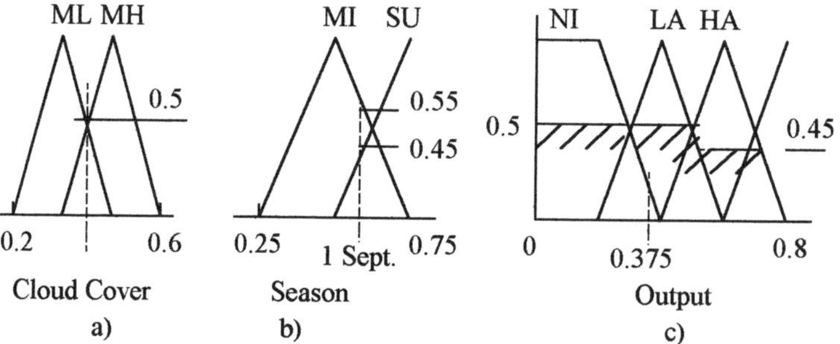

Figure Ex 3.8 Solar Collector Input and Output Data

Applying the rule-base to obtain the output inference

IF CC	AND SE		THEN OP	MIN	OUTPUT
ML	MI	LA		0.5,0.55	0.5 LA
ML	SU	HA		0.5,0.45	0.45 HA
MH	MI	NI		0.5,0.55	0.5 NI
MH	SU	LA		0.5,0.45	0.45 LA

The union of the partial inferences is shown in Figure Ex 3.9c). Defuzzifying the output by the centroid method to produce a singleton value

$$OP = \Sigma A_i L_i / \Sigma A_i = 1.8736 \text{ on the scale 0-5}$$

The customary representation is 1.8736/5 = 0.3747 on the normalised scale 0-1.

The physical output without an allowance for site factors is

$$OP' = 0.3747*18.86 = 7.067 \text{ MJ/day}$$

With a 20% reduction due to site factors, the output estimated for 01 September is; OP'' = 5.653 MJ/day.

If there was an uncertainty factor of ± 40% in the cloud cover for 01 September, then Figure Ex 3.8a) would be amended by using a fuzzy input as shown by the dotted triangle in Figure 3.9a); the result is only slightly affected by the uncertainty in the cloud cover.

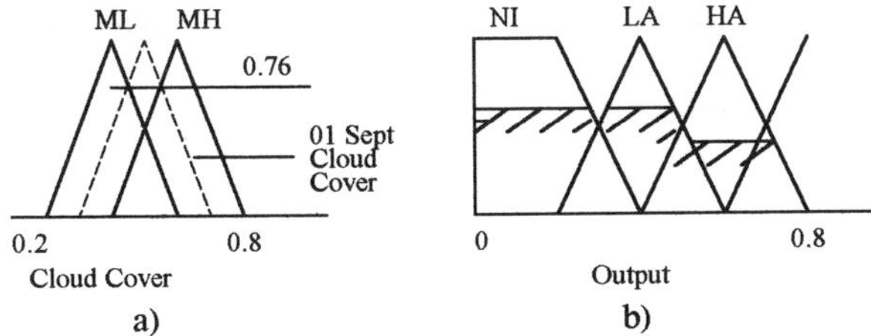

Figure Ex 3.9 Modified Diagrams for 01 September for ±40% Cloud Cover

In this case the inference is modified as follows:

IF CC	AND SE		THEN OP MIN	OUTPUT
ML	MI	LA	0.76,0.55	0.55 LA
ML	SU	HA	0.76,0.45	0.45 HA
MH	MI	NI	0.76,0.55	0.55 NI
MH	SU	LA	0.76,0.45	0.45 LA

Figure Ex 3.9b) shows thc union of the component inferences ie, the output, for this particular case.

CHAPTER 4

MATERIALS TECHNOLOGY

The traditional subject of Strength of Materials has, in more recent times, greatly increased in scope and depth in both the theoretical and experimental areas and has come to be known as Solid Mechanics or the Mechanics of Engineering Materials. Small elastic strain theory has advanced as a refined specialist study under the title of the Theory of Elasticity then later, large inelastic strain theory advanced as the Theory of Plasticity. The Theory of Elasticity exclusively considers the structural applications of materials, whilst the Theory of Plasticity considers both structural and processing applications, though the former applications are fewer than in the elastic case. Finite strain elastic theory has also been developing both with and without time delay.

From the structural viewpoint it is important to know a range of material characteristics under various environmental conditions including the following:

i) The limits of sustainable load under simple and complex stress patterns in the presence of stress concentration and cracks.
ii) The creep behaviour under sustained loads.
iii) The energy absorption capability.
iv) Fatigue resistance.
v) Fracture criteria.

The combination of the above factors in practice with the great variety of materials available in different processed and environmental conditions presents a formidable array of data, not all of which are fully understood at the present time. Even something as apparently simple as the yield stress of a material can be quite uncertain in reality, as a unique value. It is in such an environment of uncertainty or vagueness that FL has a unique role. A thorough FL treatment obviates the need to incorporate a factor of safety at the end of a design study to accommodate uncertainty in the real case because this is incorporated in the choice of fuzzy sets at the begining to represent the input data (antecedents).

In the processing of materials there is an equivalent range of factors:

i) Material flow properties under simple and complex stress fields, as a function of time temperature and deformation rate.
ii) Conditions at the tool-material interface.
iii) Processing conditions.
iv) Post-processing material state, particularly the microstructure.

Again, the practical combination of the many factors, as mentioned above for structural considerations, presents a formidable task for mathematical modelling, even if the factors were all perfectly defined, which in many cases they are not. The discussion in this chapter points to some of the critical areas in the structural application and in the processing of materials through a selection of generic cases. FL offers a deeper insight and reliability of conclusions, and also a higher integrity of results than with a purely deterministic treatment.

4.1 Elastic bending

The bending of machine, plant and structural members is one of the most familiar cases of structural loading in practice and it is of basic importance in design studies, consequently it is a topic which is often introduced in the early stages of engineering education. In initial studies the principle of St. Venant has an important explicit or implicit role, because by its application, considerable simplification of theoretical models is achieved in many important practical cases. This principle states that a given system of external forces acting on a small surface of an elastic body may be replaced by a statically equivalent system of forces acting on the same small surface area, without affecting the resulting stress fields in the body at points which are remote from the small surface area. The definition of the term "remote" in this context means a point which is unaffected by the alternative loading patterns. The argument is therefore somewhat circular. But by experimental means, such as photoelasticity, one can detect that such points do exist in sufficiently large loaded bodies. It is analogous in some respects to the boundary layer concept in fluid mechanics, but preceded it by many years.

In the analysis of practical problems, the real surface conditions at the contact points where the forces are applied, which may not be amenable to idealised modelling, are replaced by mathematical idealisations which resemble to a greater or lesser extent the actual conditions when viewed from a distance. The principle is most suited to hard elastic solids.

In the elementary theory, highly idealised mathematical models are postulated with representative loading patterns that are statically similar to simple cases found in practice. Approximate stress fields are sought through equilibrium considerations. The case of the simply supported beam with a centrally placed transverse point load is a generic problem in structural studies and may be used as an illustration of the various levels of model refinement in portraying reality. The end result of analysis is still somewhat idealised in that the load contact patterns are convenient mathematical concepts. Numerical methods remove some of the restrictions of mathematical convenience but are not a guarantee that reality will always match the analysis. The processing history of structural members will induce such phenomena as undefined anisotropy in the material which may not be uniform, and residual stresses, this will immediately make a distinction with isotropic theory and render the correspondence with reality less certain than might sometimes be supposed.

Consider the transverse loaded simply supported beam case with a rectangular cross-section, carrying a point load at mid-span, as shown in Figure 4.1 below.

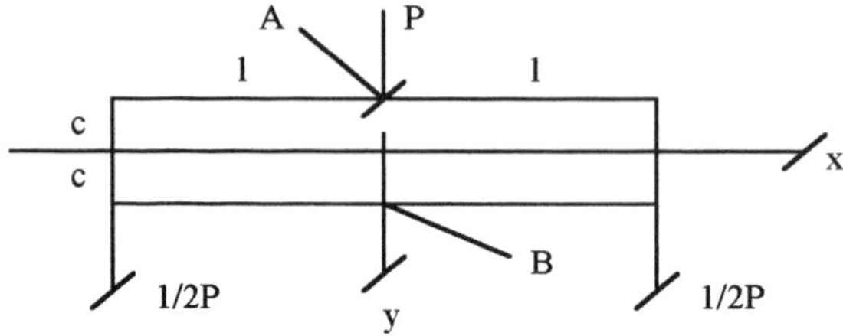

Figure 4.1 Transverse Loaded Simply Supported Beam

With the usual idealisations, as described in standard texts, the distribution of normal stress on any transverse section, at a distance of y from the neutral axis, O-x, is given by

$$\sigma_o = 3P(l\text{-}x)y/2c^3 \tag{4.1)}$$

where P is the load per unit width of the beam.

The corresponding shear stress distribution across the section is given by,

$$\tau_o = 3P(c^2\text{-}y^2)8c^3 \tag{4.2)}$$

All other stresses are assumed to be zero.

The two formulae, 4.1) and 4.2) are used extensively in the application of elementary theory and may be called the zeroth approximations. The results are satisfactory for slender beams ($c/l<<1$) of constant cross-section, but only at material points which are "distant" from the points of application of the external forces. More advanced theories focus on the conditions in the vicinity of the applied forces, but only for idealised load contact patterns. These studies yield first and second order approximations as well as an "exact" solution for the distribution of the shear and normal stress distributions in idealised hard elastic bodies.

An example of the corrections to the zeroth normal stress,σ_o, at the surface of the mid-span, point B in Figure 4.1, obtained from the various treatments taking into consideration increasing levels complexity, are shown below in table 4.1.

Table 4.1 Normal Stress Corrections at Mid-Span

Level	**Correction**
1st Approximation	-0.318P/c
2nd Approximation	-0.254P/c
Exact	-0.133P/c

The shear stress zeroth approximation, equation 4.2), is independent of the x-co-ordinate, but in the exact theory there is a load proximity correction which is a function of the x-coordinate. Also in the zeroth order theory, transverse normal stresses are ignored, whilst in the second-order and exact theories the (compressive) transverse stresses are accommodated. The theoretical surface force contact conditions are chosen on mathematical grounds and do not generally represent real loading conditions exactly. The so-called exact theory also involves rather unwieldy Fourier integrals. The same modelling choices must be made in the case of numerical analysis, of which finite difference, finite element and boundary element are the most widely used to date.

It is clear from the above discussion that in most cases the theoretical stress distribution, however derived, yields an apparent stress. In practice this is used to establish safe loading limits for design purposes based on laboratory tests of material specimens. The conditions of the laboratory test may or may not replicate the service environment exactly, also the post-processing condition of the test piece may or may not exactly replicate the condition of the actual structural component. There is also the problem of the nature of the failure and exact evaluation of the failure criterion to be used. In practice, the critical apparent stress levels in structural members are estimated by some means and compared with laboratory material tests. The next stage is specimen component tests, if this is possible, and finally knowledge of the history of the performance of similar cases completes the evidence to judge the structural reliability of a particular design in a given service environment. Thus the mathematical or numerical evaluation of apparent stresses is just one piece of the total evidence profile.

A natural way of merging the various factors which are outlined above, in a structural design study is by the application of FL methods and language. FL may not only feature in the overall evaluation, but also in the assessment of factors such as apparent stress levels

An example of a rule-base to adjust the nominal stress levels in beam bending is shown in Table 4.2. The rule-base should encapsulate expert knowledge and opinion of stress levels for a given geometry and load conditions.

Table 4.2 Rule-Base for Stress levels

		X			
Y	ZE	SM	ME	LM	LA
LN		MP	SP	ZE	ZE
MN		NE	SN	ZE	ZE
ZE	SP	ZE	SN	SN	ZE
MP	SN	ZE	SP	ZE	ZE
LP	SN	ZE	SP	SP	ZE

In the above Table, X = x/c and Y = y/c. Also ZE = zero, SM = small medium, ME = medium, LM = large medium and LA = large. Also LN = large negative, MN = medium negative, ZE = zero, MP = medium positive and LP = large positive. The

Table entries represent linguistic terms for the normal stress correction; SN = small negative, ZE = zero and SP = small positive.

For the smallest values of X and Y the normal stress correction would generally change very rapidly therefore the top left-hand corner of the above Table would require more detailed treatment.

Size effects

An interesting and important phenomenon which is not apparent in macro-stress analysis is that of size effect, whereby it is found experimentally that similar metallic specimens tend to fail at smaller values of apparent stress the larger the specimen becomes. A similar effect may also be observed with other hard elastic materials which are not necessarily brittle, but it is in some brittle materials, such as cast-iron, that it is most marked. There can be serious consequences because relatively small laboratory test pieces may show plastic flow before failure, whereas large structures of the same material may show brittle failure at smaller apparent stresses.

Example 4.1

A tapered beam is to be produced for a machine and is to carry a maximum load of 5.75 kN at a distance of 1 m from the fulcrum, as shown in Figure Ex. 4.1, the load may be suddenly applied. Due to the shape of the beam and the pivot bearing hole it is likely that the stress concentration factor at point A could be 1.14. The bearing requirements at the pivot indicate a required bearing length at A of 40 mm.

The beam is to be cast-iron of a selected grade allowing for stress concentration and size effects. Evidence of past experience has enabled a rule-base for recommended cast-iron grades to be compiled as shown in Table Ex. 4.1, together with partitioning of the universes of discourse of the stress (s), the section thickness (t) and the cast-iron grade.

Decide on the most suitable cast-iron grade to be used for this application.

Table Ex. 4.1 Stress Fuzzy Sets

		s MN/m^2					
Set	0	100	150	200	250	300	350
S_1	1	1	0	0	0	0	0
S_2	0	0	1	0	0	0	0
S_3	0	0	0	1	0	0	0
S_4	0	0	0	0	1	0	0
S_5	0	0	0	0	0	1	0
S_6	0	0	0	0	0	0	1

Table Ex. 4.2 Cross-Section Thickness Fuzzy Sets

		t mm				
Set	0	10	25	50	75	100
T_1	1	1	0	0	0	0
T_2	0	0	1	0	0	0
T_3	0	0	0	1	0	0
T_4	0	0	0	0	1	0
T_5	0	0	0	0	0	1

Table Ex. 4.3 Cast-Iron Grades Fuzzy Sets

	Grade							
Set	1	2	3	4	5	6	7	8
G_1	1	0	0	0	0	0	0	0
G_2	0	1	0	0	0	0	0	0
G_3	0	0	1	0	0	0	0	0
G_4	0	0	0	1	0	0	0	0
G_5	0	0	0	0	1	0	0	0
G_6	0	0	0	0	0	1	0	0
G_7	0	0	0	0	0	0	1	0
G_8	0	0	0	0	0	0	0	1

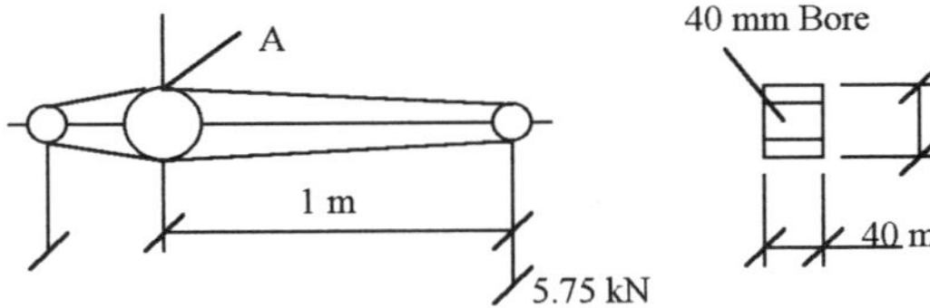

Figure Ex. 4.1 Diagram of Tapered Beam

Solution.

By simple calculation the second moment of area of the beam section at the fulcrum is

$$I = 3.12*10^{-6}m^4$$

The zeroth-order bending stress formula is

$$s_0 = by/I$$

The bending moment (b) at the fulcrum due to the load of 5.75 kN is 5.75 kNm and the distance to the outer fibres (y) at A is 50 mm.

Allowing a factor of 2 for a suddenly applied load, the bending stress is

$$s_0 = 2*5750*0.050/3.12*10^{-6} = 184 \text{ MN/m}^2$$

If there is a stress correction to the zeroth approximation of 1.14 then the estimated stress is

$$s = 1.14*184 = 210 \text{ MN/m}^2$$

To make a decision about the cast-iron grade, the antecedents (or inputs) of 210 MN/m^2 and 40 mm are applied to the tabulated fuzzy sets, Tables Ex. 4.1 and 2 respectively to find the membership values of the fuzzy sets

$s = 210$ MN/m^2	$t = 40$ mm
$\mu_{S2} = 0.58$	$\mu_{T2} = 0.4$
$\mu_{S4} = 0.42$	$\mu_{T3} = 0.6$

The FL proposition is

IF S AND T THEN G

Applying this proposition, a set of partial conclusions is found as follows:

IF S	AND T	THEN G	MIN	CONCLUSION
S_3	T_2	G_4	0.58,0.6	0.58 G_4
S_3	T_3	G_5	0.58,0.4	0.4 G_5
S_4	T_2	G_6	0.42,0.6	0.42 G_6
S_4	T_3	G_6	0.42,0.4	0.4 G_6

The above partial conclusions are aggregated to find the overall conclusion using the logic union operation, thus the decision about the cast-iron grade is

$$G = 0.58G_4 \cup 0.4G_5 \cup 0.42G_6 \cup 0.4G_6$$

i.e. $$G = 0.58G_4 \cup 0.4G_5 \cup 0.42G_6$$

The overall conclusion is shown in Figure Ex. 4.2

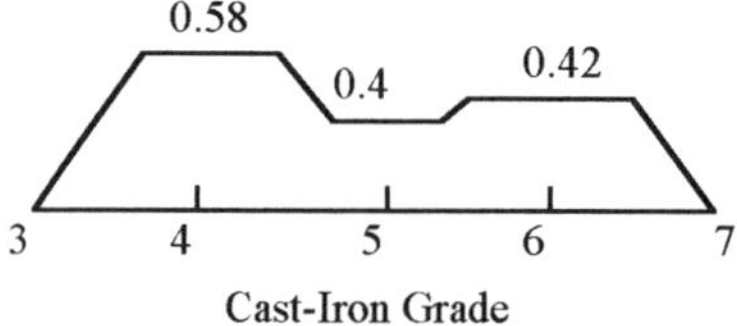

Figure Ex. 4.2 The Fuzzy Conclusion of the Cast-Iron Grade

Defuzzifying the conclusion in Figure Ex. 4.2 by the centroid method (see the Appendix) gives a value of 4.89 and when this is compared with the fuzzy sets of Grades in Table 4.3, the membership values of the cast-iron grades is $0.89G_5$ and $0.11G_4$. This is clearly weighted in favour of the G_5 grade.

In the case of flake graphite cast-iron, British Standard BS 1452:1990 specifies seven grades corresponding with G_1 to G_7 above. In the case of G_5, this would equate to grade 250 with a tensile strength of 250 MPa for a 30 mm diameter sand cast bar test piece. The full range of grades is: 100,150, 180, 200, 250, 300 and 350, representing the tensile strengths in MPa. There are similar gradings and strengths for other types of cast-irons.

4.2 Elastic instability

Columns and struts are structural members subject to compressive loads that form a generic case in instability studies and are frequently found in all types of machine, plant and frame units. The most familiar type of instability is that of buckling of long slender members when the applied compressive end load is increased beyond a certain point. In this case the main parameters of the system are the slenderness ratio, l/d, where l is the effective length of the strut between end fixings and d is some characteristic cross-section dimension, and the type of end fixing. The classical deterministic analysis for this type of elastic instability is that due to Euler, which is described in many standard texts on solid mechanics.

There are generally considered to be three classes of slenderness ratio:

i) l/d greater than about 35. Euler's theory is considered to give a reasonably good estimate of the buckling load in this case.

ii) l/d in the region of 25-30. Euler's theory becomes increasingly inaccurate as l/d is reduced in this range. Structural failure has an increasing tendency to be due to crushing, depending upon the type of material.

iii) l/d below about 25. The material tends to ductile failure or brittle fracture, depending on its type.

End fixings are normally classified into one of four groups:

a) Rounded or pin-jointed at both ends, where there is no resistance to an externally applied end couple.

b) One end encastre, the other end rounded or pin-jointed and therefore unable to resist bending in at least one plane, but free to sway sideways.

c) One end encastre, the other end rounded or pin-jointed, but restrained from side-sway.

d) Both ends encastre and no side sway.

These are ideal end conditions and in practice it is often found that they are only partially satisfied and may comprise a mixture of the ideals. It is under uncertainty such as this that the FL type of treatment becomes appropriate.

The theory of Euler considers the case of a long slender column or strut with rounded or pin-jointed ends and direct stress on the cross-section is neglected. The theory predicts a buckling load,P_e, given by

$$P_e = \pi^2 EI/l^2 \tag{4.3}$$

where E is Young's modulus of the material, I is the cross-section second moment of area about an axis perpendicular to the plane of buckling and l is the effective length of the column. For the four end conditions; a) to d) above, the critical load may be represented in terms of the Euler load by

$$P = KP_e \tag{4.4}$$

The values of K are given by

Case	a)	b)	c)	d)
K	1	1/4	2	4

For short and intermediate length columns the direct stress cannot be ignored (by definition). Both these cases have been the subject of further analysis for both ductile yielding, which is important in the processing of materials, and the practically important intermediate case. Several formulae have been proposed for these cases and may be found in texts treating elastic or plastic strain. In the case of structural members it is normal to include a factor of safety to accommodate unmodelled factors.

Here again there is a degree of uncertainty, partly dependent upon the environmental conditions (such as temperature) and partly upon the material type.

Such factors as those discussed above may be collected together and evaluated in a rule-base as shown in the example in Table 4.3. below; where P_c is the failure load in compression for short columns.

Table 4.3 Rule-Base for P/P_c

	l/d		
K	SH	LM	LO
A	HM	LM	LO
B	HI	LM	LO
C	HI	HM	HM
D	HI	HI	HI

In the above Table, SH = short, ME = medium, LG = long. Also LO = low load, LM = low medium load, HM = high medium load, HI = high load.

Example 4.2

A column has an effective *l*/d ratio of about 10 and end restraint conditions that would give a K factor of about 1.5, which means that the column is partially restrained at the free end and is mid-way between cases b) and c). The partitioning of the universes of discourse is shown below in Tables Ex. 4.4., 5 and 6.

Table Ex. 4.4 Slenderness Ratio, *l*/d, Fuzzy Sets

	***l*/d**		
Set	1	30	100
SH	1	0	0
ME	0	1	0
LE	0	0	1

Table Ex. 4.5 K Factor Fuzzy Sets

	K				
Set	0	0.25	1.0	2.0	4.0
A	1	1	0	0	0
B	0	0	1	0	0
C	0	0	0	1	0
D	0	0	0	0	1

Table Ex. 4.6 P/P_c Fuzzy Sets

	P/P_c				
Set	0	0.55	0.7	0.85	1.0
LO	1	1	0	0	0
LM	0	0	1	0	0
HM	0	0	0	1	0
HI	0	0	0	0	1

Assume that the rule-base given in Table 4.3 applies.

Solution

The antecedents (inputs) given are; $l/d = 10$ and K factor = 1.5. Using these values, the membership values may be obtained by interpolation in Tables 4.4 and 5 respectively. The resulting values are as follows:

$l/d = 10$	$K = 1.5$
$\mu_{SH} = 0.35$	$\mu_B = 0.5$
$\mu_{ME} = 0.65$	$\mu_c = 0.5$

The FL proposition is

IF *L/D*	AND K	THEN P/P_C	MIN	CONCLUSION
ME	B	LM	0.65,0.5	0.5 LM
ME	C	HM	0.65,0.5	0.5 HM
SH	B	HI	0.35,0.5	0.35 HI
SH	C	HI	0.35,0.5	0.35 HI

The partial solutions are aggregated by the logic union operation to give the fuzzy buckling load ratio P/P_c shown in Figure Ex. 4.3.

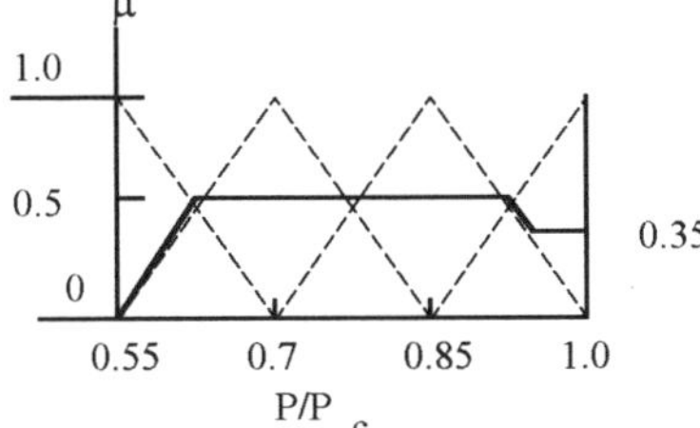

Figure Ex. 4.3 Fuzzy Failure Load Ratio, P/P_c

With the data given Figure Ex. 4.3 concludes quite a wide spread of possible failure load ratio. The source of this is the coarse partitioning of the slenderness ratio in Table Ex. 4.4, and a small number of rules in the rule-base, Table 4.3. A more extensive knowledge of the physical properties of the material as a column would enable the rule-base to be extended with higher definition. This would consequently give higher definition in Figure Ex. 4.3.

The above example is the simplest case of the failure of a column or strut. There are other cases of practical importance such as thin walled columns that "crinkle", columns with eccentric loading, initially curved columns and also the effects of various types of lateral loading, to name some but not all of the possibilities. Another advantage of the FL treatment is that if the nature of the end constraint is uncertain then they may be represented as a fuzzy sets, for example the K factor might be

$$K = [0//1.25+1.0//1.5+0.5//1.75+0//2.0] \quad 4.5)$$

The intersection of this with the partitioning fuzzy sets in Table Ex. 4.5 then gives the membership values (μ).

The crushing load of a short column of the same material with the same cross-section may not be clearly defined due to a number of factors:

i) Friction at the platen-specimen interface
ii) Specimen barrelling
iii) Definition of failure
iv) Complex stress distribution throughout the specimen
v) The state of the material

Some of these factors would also affect the behaviour of intermediate length specimens.

4.3 Failure of an assembly

An assembly of component parts comprising a complete structure may be subject to a variety of loading patterns and the resulting various possible modes of failure are of importance in assessing the different levels of safety or risk. There are also tolerances of manufacture and composition which create a level of indeterminacy and which affects the precision of knowledge of the physical conditions within a system. A FL treatment is well suited to such circumstances and is able to express imprecise knowledge and to provide valid operations by which reasonable deductions may be made. These factors may be expressed by the use of fuzzy sets rather than deterministic values, as described before. Attention here is directed towards the effect od load patterns on the safety of assemblies.

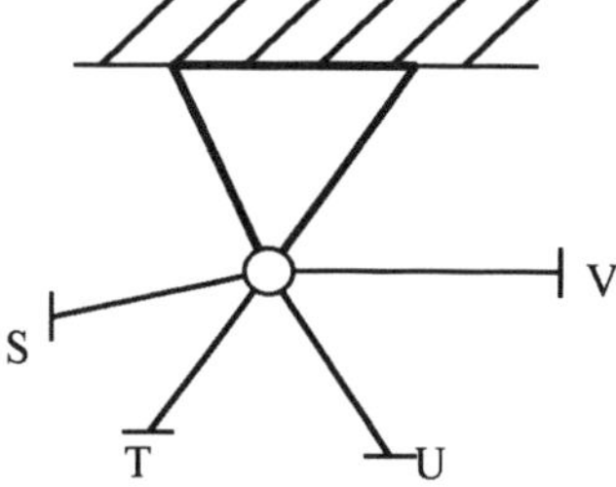

Figure 4.2 Load Combination

Consider, for example, a simple pin-jointed triangular frame as shown in Figure 4.2. with the load combination S, T, U and V. Suppose from tests or other information it is established that a safe combination of these is given by

$$C = [\mu_S//S+\mu_T//T+\mu_U//U+\mu_V//V] \qquad 4.6)$$

arranged as a convex set.

Now let a possible applied load pattern be given by

$$A = [\mu_K//K+\mu_L//L+\mu_M//M+\mu_N//N] \qquad 4.7)$$

also arranged as a convex set.

Various consequences may be deduced by the use of fuzzy logic operations as follows:

a) The set of loads for which the assembly is safe is $Y_S = C \cap A$. (1,2,5,3)

b) The set of loads for which the assembly is unsafe is $Y_U = C' \cap A$. (1,4,5,6,7,8)

c) The set of loads for which the assembly is safe and unsafe is $Y_{SU} = C \cap C' \cap A$. (1,2,5,6,7,8)

The above are illustrated in Figure 4.3 below.

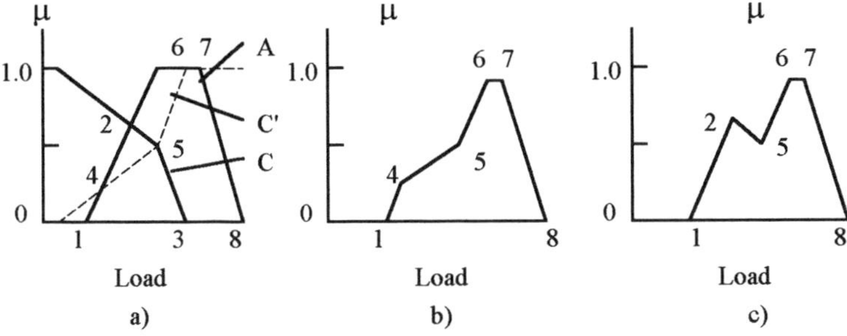

Figure 4.3 Safe and Unsafe Assembly Sets

Example 4.3

An assembly comprises a pin-jointed fork and eye which is subjected to an applied axial load L. In destructive test of the assembly there are three possible failure modes:

i) The pin may fail.
ii) The eye may fail.
iii) The fork end may fail.

It is found from a number of repetitive test that the fork end and the eye are the most likely to fail and that the pin is relatively safe. From the tests test results it is established that the safe loads for the fork
and eye are F and E respectively where F and E are given by

$$F = [1.0//0+1.0//8+0.5//20+0//25] \text{ kN} \quad E = [1.0//0+1.0//10+0.5//15+0//20] \text{ kN}$$

The applied load is most likely to have the following profile

$$L = [0//15+1.0//20+0//30] \text{ kN}$$

Estimate the loads for the following conditions:

i) Both fork and eye are safe.
ii) The eye is safe but the fork is not.
iii) The eye and fork are both unsafe.

Solution

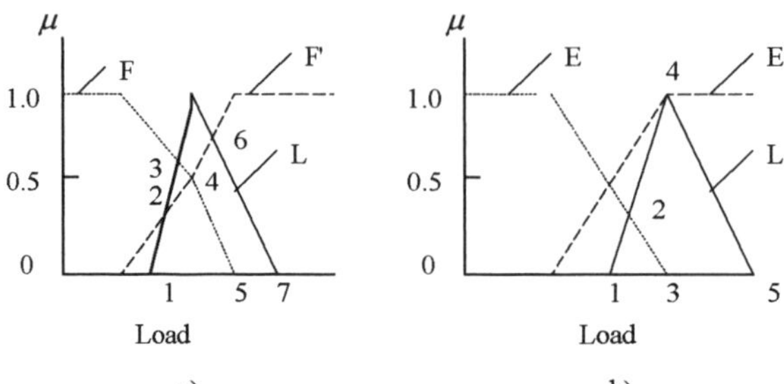

Figure Ex. 4.4 Applied Load, Fork End and Eye Failure Loads
a) Fork End Safe Load (F) and Failure Load (F')
b) Eye Safe Load (E) and Failure Load (E')

Figure Ex. 4.4 a) and b) illustrate the applied load set, the forkend and eye safe and failure load sets. In a) the outline 1,2,3,4,5 represents the safe applied load profile for the fork end, whilst the outline 1,2,6,7 represents the unsafe applied load profile. In b) the outline 1,2,3 represents the safe applied load profile for the eye end , whilst the outline 1,4,5 represents the unsafe applied load profile. It is clear that the b) 1,2,3 profile is contained in the a) 1,2,4,5 profile.

Conclusions:

i) The b) subset 1,2,3 represents the load condition for which both fork end and eye end are safe.
ii) The eye end is safe for the b) subset 1,2,3 whilst the fork end is unsafe for the a) subset 1,2,4,6,7. Therefore the eye end is safe and the fork end is unsafe for the b) subset 1,2,3.
iii) The eye end is unsafe for the b) subset 1,2,4,5, whilst the fork end is unsafe for the a) subset 1,2,4,6,7. Therefore they are both unsafe for the a) subset 1,2,4,6,7.

It will be noted that in FL a component can be both safe and unsafe. The reason for this is that in FL the union of a set and its complement does not result in a null set.

4.4 Polymeric solids

Manufactured polymeric solids containing fillers, reinforcing fibres, plasticisers and pigments in varying proportions, depending upon the application and are customarily known as plastics. They usually fall into either the category of thermosetting resins or thermoplastics which are reversibly softened by the application of heat. The thermosetting plastics, such as the epoxy resins, phenol formaldehyde and some polyesters, for example, are heavily cross-linked in their molecular structure and do not soften much on the application of heat, even up to the point of decomposition, and they consequently cannot be reprocessed. The thermoplastic group however soften from an initial glassy state at low temperatures to a rubbery state above the so-called glass transition temperature and finally to a viscoelastic polymer melt before the decomposition temperature is reached. A noteable feature of polymers is their delayed response to applied stress. A feature of a true solid is that it has a reference configuration of permanent significance; not all plastics have this feature and incomplete (delayed) deformation recovery is found in some after removal of the load.

The average molecular weight of commercial plastics may range up to about 10^6 in an amorphous or crystalline structure, or a combination of both. The molecules may be linear or branched and may be cross-linked in varying degrees to form a three-dimensional network. The rheology of plastics has been extensively investigated and there is a considerable body of published work available on the stress-strain-time-temperature relationships of these materials. The mechanical properties of plastics may range from linear with small strains to non-linear with large strains; this aspect is considered below. Some plastics are also noticeably affected by environmental conditions such as humidity and solar radiation which can affect the mechanical and other properties.

Creep

A significant feature of thermoplastics is their tendency to creep in time under an applied external load, or due to residual processing stresses. This becomes much more apparent as the temperature rises. For small strains the strain, $\varepsilon(t)$, at a particular time t arising from a stress history $\sigma(t')$ may be expressed in integral form as

$$\varepsilon(t) = \int J(t-t')(d\sigma(t')/dt')dt' \qquad \text{(limits: -infinity,t)} \qquad 4.8)$$

where J(t-t') is called the creep compliance. Equation 4.8) depends upon the validity of the linear superposition principle.

To completely characterise the linear rheological behaviour of a material under isothermal conditions the bulk modulus (K) or the Poisson's ratio (ν) is also required. The former is convenient in practice as volumetric strains are usually smaller than the shear strains. The shear modulus would typically be about 1.0 GPa for small strains. and the bulk modulus about 4.5 GPa.For large shear strains the material may be often considered as incompressible. Normally, volumetric strains of solid polymers never become large, whereas shear strains may do so in thermoplastic materials.

The creep compliance of amorphous polymers may increase by a factor of 10^3 to 10^4 over a small temperature rise around the glass transition temperature. Above the glass transition temperature it may be possible to use a time-temperature equivalence relationship to obtain superposition of material properties at different times and temperatures.

An idealised impression of how plastics behave under load may be obtained by considering some basic mathematical models. Linear and non-linear viscoelastic rheological properties are considered below; both of these may be found in varying proportions in a plastic material.

Linear Viscoelasticity

A simple model of this type is given by the addition of an elastic and a viscous stress response

$$\sigma = E\varepsilon + \mu d\varepsilon/dt \qquad 4.9)$$

or,

$$\sigma/E = \sigma J = \varepsilon + \lambda d\varepsilon/dt \qquad 4.10)$$

where, $\lambda = \mu/G$; the relaxation time and J is the shear compliance.

This model is only able to represent the mechanical response of plastics in a gross way. By standard mathematical manipulation, equation 4.10) may be transformed into an integral form by using an integrating constant. Following an arbitrary applied stress history, $\sigma(t')$, the strain at current time is given by

$$\varepsilon(t) = J/\lambda \int \sigma(t')\exp\text{-}(t-t')/\lambda dt' \quad \text{(limits: -infinity,t)} \qquad 4.11)$$

Real materials actually show a spectrum of relaxation times and therefore a more general representation of the response is given by

$$\varepsilon(t) = \int N(\lambda)/\lambda \int \sigma(t')\exp\text{-}(t-t')/\lambda dt'd\lambda \qquad 4.12)$$

where $N(\lambda)$ is called the relaxation spectrum, integrated over zero to infinity. Equation 4.12) is a refinement of the simple model given by equation 4.10).

Suppose that a constant stress σ_0 is suddenly applied to the material at time $t' = t_0$, then $\sigma(t') = \sigma_0 H(t-t_0)$, where $H(t-t_0)$ is Heaviside's unit step function. The current compliance is then given by

$$J(t) = \varepsilon(t)/\sigma_0 = \int N(\lambda) \int_{t_0}^{t} 1/\lambda \exp\text{-}(t-t')/\lambda dt'd\lambda \qquad 4.13)$$

Integrating with respect to temporal time yields

$$J(t) = \int N(\lambda)(1-\exp\text{-}(t-t_0)/\lambda)d\lambda \qquad 4.14)$$

The simplest type of relaxation spectrum is that where there are only two discrete relaxation times, one of which is zero and the other is finite but non-zero. The relaxation spectrum is then given by

$$N(\lambda) = J_0\delta(\lambda) + J_1\delta(\lambda-\lambda_1) \qquad 4.15)$$

where $\delta(\ )$ is Dirac's delta function.

Substituting the distribution given in equation 4.14) into equation 4.13) and integrating gives

$$J(t) = J_0 - J_1 \exp-(t-t_0)/\lambda_1 \qquad 4.16)$$

For very small values of $(t-t_0)$

$$J(t_0) = J_0 - J_1 \qquad 4.17)$$

This is the instantaneous elastic compliance. Also for very large values of $(t-t_0)$, J( ) = J_0. This is the long term elastic compliance.

The above results are only realistic for infinitesimally small strains.

Non-Linear Viscoelasticity

In all plastics except those in the hard elastic state, finite strains may be caused either by the action of long term applied loads or through the effect of residual processing stresses. In this case two important features emerge:

i) The shear stress-shear strain relations become non-linear.
ii) In simple shearing, normal stresses are generated which are an even function of the shear strain.

In some of the engineering literature on the mechanical properties of plastics, item i) is recognised but item ii) is not. A more thorough treatment of the above effects would require a tensor representation of the rheological relations, the equations of equilibrium and the compatibility equations. As an alternative the literature adopts an empirical approach, based on the results of simple tensile loading (the mode of much laboratory testing), or simple shear, without normal stress effects measurement. Interpreting the results an assumption is often made that the rheological relationship may be expressed as functional products of time and stress, say

$$\varepsilon(t,\sigma) = \int f(t-t')g(\sigma)dt' \qquad \text{(limits: -infinity,t)} \qquad 4.18)$$

The development of this formula is facilitated by assuming a power-law dependency for both thc f( ) and g( ) functions, for example

$$f(t-t') = a_0\delta(t-t') + a_1(t-t')^n \qquad (0<n<1) \qquad 4.19)$$

and $$g(\sigma) = b\sigma^{m+1} \qquad (0<m<1) \qquad 4.20)$$

a_0,a_1 and b are material properties that depend only upon temperature.

As an illustration of the material response portrayed by the above type of model, consider again the case of a stress suddenly applied at time t' = t_0, then

$$\varepsilon(t) = b\sigma_0^{m+1} \int \{a_0\delta(t-t') + a_1(t-t')^n\}dt' \qquad \text{(limits, } t_o \text{ to t)} \qquad 4.21)$$

or, $$\varepsilon(t) = b\sigma_0^{m+1}\{a_0 + a_1(t-t_0)^n/(n+1)\} \qquad 4.22)$$

Consequently, the non-linear compliance, L(t), is given by

$$L(t) = \varepsilon(t)/\sigma_0 = b\sigma_0^m\{a_0+a_1(t-t_0)^n/(n+1)\} \qquad 4.23)$$

Such a treatment may be useful for simple tension, compression or shear in design work, but as indicated before, it does not account for the proper tensor nature of the material properties that would be important under finite strains, and hence would be unsuitable to treat a more complex stress field. It will also be recalled that polymer processing will generally impart significant directional properties to the material and there would be significant residual stresses in the material.

Summarising the above outline of rheological properties of plastics, it may be stated that normally the material will exhibit the following types of rheological properties in varying degrees:

i) Linear elastic (LE): Small strains, when the time of application of the load is much less than the characteristic time delay of the material.
ii) Linear viscoelastic (LV): Small strains, when the time of application of the load is of the same order as the characteristic time delay of the material.
iii) Non-linear viscoelastic (NV). Finite strains, when the time of application of the load is the same order or larger than the characteristic time delay of the material. The temperature is greater than the glass-transition temperature.

FL treatment

If the mathematical models of the material are adequate for design considerations then there is no need to consider fuzzy relations for mechanical properties. But if the service conditions such as stress level, temperature or length of time of the load application are such that more than one range of material response is spanned, then a FL treatment becomes appropriate. The categories of response of a particular material are not sharply defined, but merge into adjacent categories in a fuzzy way.

It is convenient to work in terms of dimensionless quantities. Let the dimensionless stress, temperature and time be; ϕ,ζ and ξ respectively, then

$$\phi = \sigma/\sigma';\zeta = (T-T')/(T''-T');\xi = t/\lambda$$

where σ' is a characteristic stress, such as the failure stress, T' is a reference temperature, T'' is the melt temperature and λ is the characteristic time-delay of the material.

A typical rule-base relating the categories of mechanical response to the applied stress and temperature would be as shown in Table 4.4.

Table 4.4 Categories of Mechanical Response

ζ \ ϕ	ZE	SM	LA
LO	LE	LE	VE
ME	LE	VE	NV
HI	VE	NV	NV

The specific form of the rule would depend upon the type of plastic. The linguistic variable entries in the rule-base refer to compliance formulae as follows:

$$LE: J_{LE} = J_0 - J_1 \quad 4.24)$$

$$VE: J_{VE} = J_0 - J_1 \exp-(t-t_0)/\lambda_1 \quad 4.25)$$

$$NV: J_{NV} = b\sigma_0^m\{a_0 - a_1(t-t_0)^n/(n+1)\} \quad 4.26)$$

Example 4.4

A plastic component is expected to sustain a constant in-service tensile load producing a nominal stress of 2 MPa. It is required to estimate the associated direct strain after a period of 9 weeks has elapsed. The material has been subjected to laboratory tests from which it is known that at the service temperature, the mechanical properties are

Instantaneous elastic compliance	1.0 GPa	
Total long-term compliance	5.0 GPa	
Non-linear compliance	$J_{NV} = \sigma^{0.5}\{0.7071+1.932*10^{-4}t^{0.6}\}$ MPa	
Relaxation time	$9*10^6$ seconds	
Reference stress σ'	3.57 MPa	(Yield stress)
Dimensionless service temperature	0.44	

The rule-base is assumed to be as shown in Table 4.4 and the partitioning of the stress and temperature fields is shown below in Table Ex. 4.7 and Table Ex. 4.8.

Table Ex. 4.7 Stress Fuzzy Sets

Set \ ϕ	0	0.25	0.5	0.75	1.0
ZE	1	1	0	0	0
SM	0	0	1	0	0
LA	0	0	0	1	1

Table Ex. 4.8 Temperature Fuzzy Sets

Set \ ζ	0	0.25	0.5	0.75	1.0
LO	1	1	0	0	0
ME	0	0	1	0	0
MI	0	0	0	1	1

Estimate the effective compliance at the end of the nine week period.

Solution

The dimensionless applied stress is, $\phi = \sigma/\sigma' = 2/3.57 = 0.56$
The instantaneous elastic compliance $(J_0\text{-}J_1) = 1.0\ \text{GPa}^{-1}$

After 9 weeks ($5.443*10^6$seconds) the delayed linear viscoelastic compliance is given by

$$J_{VE}(t) = J_0\text{-}J_1\exp\text{-}t/\lambda$$

or $$J_{VE}(9) = 5\text{-}4\exp\text{-}(5.443/9) = 2.805\ \text{GPa}^{-1}$$

After 9 weeks the delayed non-linear viscoelastic compliance is given by

$$J_{NV}(9) = \{0.7071+1.932*10^{-4}(5.443*10^6)^{0.6}\} = 4.006\ \text{GPa}^{-1}$$

Using Tables Ex. 4.7 and 8 the antecedents (inputs) may be transformed to fuzzy linguistic sets with membership values, with the following results:

σ	ζ
$\mu_{\Sigma M} = 0.75$	$\mu_{ME} = 0.8$
$\mu_{LA} = 0.25$	$\mu_{LO} = 0.2$

Using the results of the rule-base in Table 4.4. the following inference table may be constructed,

IF ϕ	AND ζ	THEN J	MIN	CONCLUSION
SM	ME	VE	0.75,0.8	0.75 VE
SM	LO	LE	0.75,0.2	0.2 LE
LA	ME	NV	0.25,0.8	0.25 NV
LA	LO	VE	0.25,0.2	0.2 VE

The final column on the right-hand side of the above tablulation gives the partial fuzzy conclusions. The overall conclusion represents the compliance at a time interval of 9 weeks from the initial application of the load , and it is obtained as the union of the partial conclusions, thus the compliance fuzzy set is

$$J(9) = [0.75VE+0.2LE+0.25NV]$$
$$= [0.75//1.0+0.2//2.805+0.25//4.006\}$$

A deterministic value may be estimated by defuzzifying this set (see the Appendix for methods)

$$J(9) = \Sigma\mu_iJ_i/\Sigma\mu_i = \{0.75*1.0+0.2*2.805+0.25*4.006\}/\{0.75+0.2+0.25\}$$
$$= 1.927\ \text{GPa}^{-1}$$

Although the non-linear element has a membership value of only 0.25, it does infact contribute about 52% of the final compliance value and is therefore relatively more important than the two linear elements.

4.5 Press brake bending

This is a basic and widely used generic fabrication process involving the bending of a flat sheet of metal between a vee-shaped tool and die. In smaller operations it is often manually powered. In deterministic analysis, the classical theory of bending of curved bars provides the basis of the theoretical work. The theory provides the distribution of stresses in a narrow rectangular cross-section; it is essentially a two-dimensional stress theory. The theory is also considered to be valid for the other extreme case of a sheet of metal where the dimension perpendicular to the plane of bending is large compared to the sheet thickness and is the case of plane strain. This is the normal press brake bending operation. There is also theory available for the elastoplastic bending and the fully plastic bending of wide plates.

From the theoretical viewpoint, plate bending between three rollers is related to press brake bending, though operationally the former is a continuous process, whilst the latter is essentially a batch process. Tool and die design requires some expertise even though the tool and die geometry are simple. In practice the workpiece may experience cracking on the outer surface of the corner formed by the tool and die, where the maximum tensile strain occurs. There is also normally an elastic springback of the metal after the forming operation which decreases the angle formed by the tool and die. Springback may be reduced to a certain extent by an ironing pressure applied by increasing the tool-die force on the workpiece when it is fully formed and touches the bottom of the die.

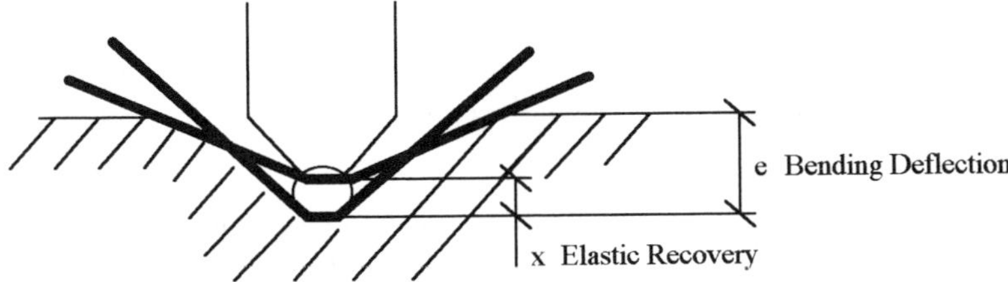

Figure 4.4 Press Brake Bending

The tool force (B) required to fully form unit width of the bend without ironing may be expressed in functional form as

$$B = f(Y,t,e,r) \quad 4.27)$$

where Y = the yield stress criterion, t = material thickness, e = the bending deflection and r = tool nose radius.

In dimensionless form, this may be expressed as

$$B' = f(e',t') \quad 4.28)$$

where $B' = B/Yt$, $e' = e/r$ and $t' = t/r$

The springback (x) without ironing pressure may similarly be expressed as

$$x = f(z,e,t,r,a) \quad 4.29)$$

where z = elastic strain recovery in a laboratory specimen and a is the reduction of area per unit in a laboratory tensile test (a measure of material ductility).

In dimensionless form, this may be expressed as

$$x' = f(z,t',a,e') \quad 4.30)$$

Let g be the elastic springback with an applied ironing pressure (w) per unit workpiece width then

$$g' = f(z,t',w',e') \quad 4.31)$$

where, g' = g/e and w' = w/Yr.

Theory indicates that the moment required to produce elastic bending in a curved plate may be approximately expressed as

$$M_c = Y\{(b^2-a^2)-4a^2b^2(\ln b/a)^2\}/(a^2-b^2+2b^2\ln b/a) \text{ per unit width.} \quad 4.32)$$

Whilst in the fully plastic state it is

$$M_P = Y(b-a)^2 \text{ per unit width} \quad 4.33)$$

where b-a = t, also a is the inner and b is the outer radius of the corner formed. This assumes no work hardening.

For a ratio of b/a = 2, then M_P/M_c is approximately 2. For the practical case however the stress fields imposed by the tool and die together with friction between them and the workpiece mean that the theoretical assumptions rapidly become invalid as the workpiece deflection increases.

The minimum allowable bend radius is related to the angle of bend and to the thickness to radius ratio. The relationship is empirical and very specific to the material, the tool and die geometry and the work conditions such as lubrication. A simple rule is that the maximum bending strain should not exceed 75% of the fracture strain, but obviously this is difficult to apply in practice. The tool force is also similarly empirically related to the other parameters. The knowledge of operationally important factors is capable of being cast into rule-based formulations which can form the basis of design and operating decisions.

An example of a rule-base relating the tool force with the depth of draw and material thickness (all dimensionless), comprising sixteen rules is given in Table 4.5.

Table 4.5 Rule-Base for Tool Force

	e'			
t'	ZE	LM	HM	HI
ZE	ZE	ZE	LO	LO
SM	ZE	LO	LO	MO
LA	LO	LO	MO	MO
VL	LO	MO	MO	LA

where;ZE = zero, LM = low medium, HM = high medium, HI = High, SM = small medium, LA = large VL = very large, LO = low and MO = moderate.

There is a progression from the least in top left-hand corner of the above array, to the greatest in the bottom right-hand corner as conditions become more onerous. The universes of discourse of e' and t' cover practical ranges of a given process plant capacity and type of material. The partitioning of the universes of discourse would be chosen to give finer divisions where greatest definition is required.

Example Ex. 4.5

A sheet metal component is to be pressed to form a corner angle bend in a press brake operation. The corner is to have an internal radius of 2.86 mm. The sheet metal is 2 mm thick and has a yield stress of 300 MPa. It is estimated that a draw depth (bending deflection) of 15.73 mm will give the required corner angle, allowing for elastic springback. Assume that the expert knowledge base is as shown in Table 4.5.

Estimate the machine capacity required per metre width of sheet.

The partitioning of the universes of discourse is as shown in Tables Ex. 4.9, 4.10 and 4.11 below.

Table 4.9 Partitioning of the Draw Depth

	e'					
Set	0	2	4	6	8	10
ZE	1	1	0	0	0	0
LM	0	0	1	0	0	0
HM	0	0	0	1	0	0
HI	0	0	0	0	1	1

Table 4.10 Partitioning of the Sheet Thickness

	t'					
Set	0	0.2	0.4	0.6	0.8	1.0
ZE	1	1	0	0	0	0
SM	0	0	1	0	0	0
LA	0	0	0	1	0	0
VL	0	0	0	0	1	1

Table 4.11 Partitioning of the Tool Force

	B'					
Set	0	0.2	0.4	0.6	0.8	1.0
ZE	1	0	0	0	0	0
LO	0	1	0	0	0	0
MO	0	0	1	0	0	0
LA	0	0	0	1	1	1

Solution

It is easily calculated that the dimensionless thickness (t') and draw depth (e') are; t' = 0.35 and e' = 5.5. Using these values and interpolating in Tables 4.9 and 10 to find the linguistic sets and their associated membership values gives

t' = 0.35	e' = 5.5
$\mu_{LM} = 0.25$	$\mu_{ZE} = 0.25$
$\mu_{HM} = 0.75$	$\mu_{SM} = 0.75$

Using a FL proposition to find partial conclusions using the rule-base gives

IF t'	AND e'	THEN B'	MIN	CONCLUSION
LM	ZE	ZE	0.25,0.25	0.25ZE
LM	SM	LO	0.25,0.75	0.25LO
HM	ZE	LO	0.75,0.25	0.25LO
HM	SM	LO	0.75,0.75	0.75 LO

The overall conclusion is found from the union of the partial conclusions in the above array, yielding

B' = [0.25ZE+0.75LO]

or B' = [0.25//0+0.75//0.2]

This may be defuzzified to give B' = 0.15

Now B' is defined as B' = B/Yt

Hence, $B = B'Yt = 0.15*300*10^6*0.002 = 90$ kN/m

In practice, the value to be assigned to the yield criterion is often ill-defined and accompanied by work-hardening. Hence this quantity would be more properly defined by a fuzzy set, say

$$Y = [0//298+0.5//324+1.0//350+0.5//376+0//403] \quad 4.34)$$

Then $$B = B' \times Y \times T \quad 4.35)$$

Elastic springback may also be treated in a similar way based upon equation 4.29) or 4.30), though this case is complicated by having four terms on the right-hand side of the equations. This requires a series of rule-bases in each of which one of the parameters is constant, and incremented for each of the Tables.

4.6 Forging and extrusion

From a theoretical point of view, forging and indentation are quite similar, extrusion too has similarities in the physical models of the process, though forging and indentation are both batch processes in which the steady state is not achieved, whereas it is often achieved in extrusion. In all three processes both hot and cold working are widely practised, though less frequently in the indentation process. The forging and extrusion processes both confer enhanced material properties and produce anisotropy in the material unless the process is carried out above the recrystallisation temperature of the metal. Extrusion is also widely used as a production process in the plastics manufacturing industry, but always above the melt temperature.

It is not uncommon to find quite a wide scatter in the results of forging experiments as a result of the effects of non-ideal conditions such as measurement errors and uncontrolled factors such as variations in the lubrication of the workpiece and tool interface, especially when comparing the results from different sources. The type of scatter which may be experienced is shown in Figure 4.6.

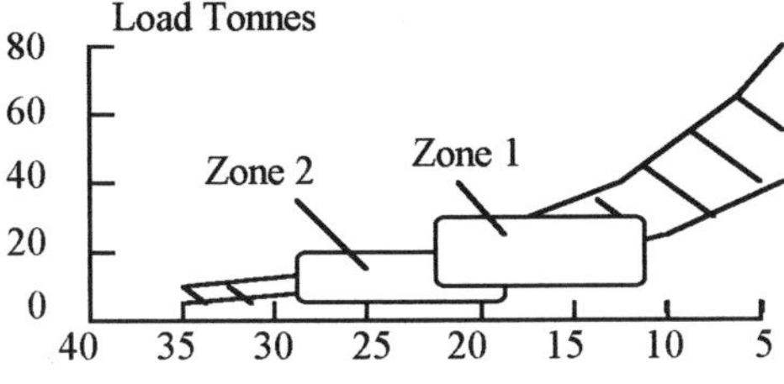

Figure 4.5 Forging Load v Web Thickness

The chart in Figure 4.5 shows the upper and lower boundaries of most of the test data for a typical forging operation in which an H section component is produced in a single operation. There is significant scatter and thus uncertainty especially at the lower end of the web thickness scale.

To interpret such data in FL terms, the data is divided into several zones, as shown in Figure 4.5. Within these zones the web thickness and forging load may be expressed as a fuzzy set pairs, one pair for each zone. The number of zones chosen will determine the coarseness of the final relationship.

The local relationship, $R_{i,}$ may be constructed by firstly finding the Cartesian product of the local fuzzy set pairs $(L_{i,}T_{i})$

$$R_i = L_i \times T_i \tag{4.36}$$

The overall relationship sought is found from the union of the R_i local relationships

$$R = R_1 \cup R_2 \cup R_3 \cup \ldots R_n \tag{4.37}$$

From the relationship various deductions may be made, such as the forging load for a desired web thickness

Example Ex. 4.6

Consider for example the data in Figure 4.6,divided into just two zones as shown. Find the forging load required for a web thick of 15 mm.

Solution

The load and the web thickness are expressed as pairs of fuzzy sets

L_1 = [0//5+0.5//7.5+1.0//10+0.5//15+0//20]
T_1 = [0//10+0.5//12.5+1.0//15+0.5//17.5+0//20]

and L_2 = [0//0+0.5//2.5+1.0//5+0.5//7.5+0//10]
T_2 = [0//10+0.5//12.5+1.0//15+0.5//17.5+0//20]

Relations between L_1 and T_1 and also between L_2 and T_2 may be found by taking their Cartesian products, thus

$R_1 = L_1xT_1$ and $R_2 = L_2xT_2$

Hence, R_1 =

	L_1 Tonnes				
T_1 mm	5	7.5	10	15	20
10	0	0	0	0	0
12.5	0	0.5	0.5	0.5	0
15	0	0.5	1.0	0.5	0
17.5	0	0.5	0.5	0.5	0
20	0	0	0	0	0

and R_2 =

	L_2 Tonnes				
T_2 mm	0	2.5	5	7.5	10
15	0	0	0	0	0
17.5	0	0.5	0.5	0.5	0
20	0	0.5	1.0	0.5	0
26	0	0.5	0.5	0.5	0
32	0	0	0	0	0

The overall relationship may then be found from the union of R_1 and R_2

$R = R_1 \cup R_2$

Substituting the above arrays into this equation gives

R =

	L Tonnes						
T mm	0	2.5	5	7.5	10	15	20
10	0	0	0	0	0	0	0
12.5	0	0	0	0.5	0.5	0.5	0
15	0	0	0	0.5	1.0	0.5	0
17.5	0	0.5	0.5	0.5	0.5	0.5	0
20	0	0.5	1.0	0.5	0.5	0.5	0
26	0	0.5	0.5	0.5	0	0	0
32	0	0	0	0	0	0	0

For each value of flange thickness there is a corresponding forging load fuzzy set. In this example the flange thickness is 15 mm. By inspection of the R array above the forging load is given by

L = [0//5+0.5//7.5+1.0//10+0.5//15+0//20]

The illustration of this fuzzy load set as a continuous distribution is shown in Figure Ex.4.12.

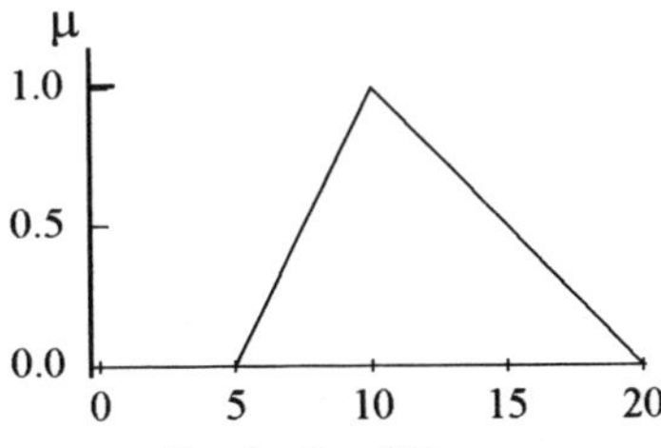

Figure Ex. 4.5 Forging Load Fuzzy Set

It may be seen that the forging load is about 10 tonnes plus. Defuzzifying the set by the centroid method gives forging load of 11.67 tonnes. Defuzzifying the discrete fuzzy load gives a value of 10.6 tonnes. The discrete fuzzy set is an approximation to the continuous set. It may be noted that the forging load fuzzy set is in agreement with Figure 4.6.

CHAPTER 5

AUTOMATIC CONTROL

The control of systems or processes is conventionally subdivided into problems of tracking and regulation. The basic tracking problem is that of causing a system to follow the movements of a target with minimum error, whilst that of regulation is to cause a process to deliver a given output whilst correcting any tendency to drift away from the set point. There are many variants of these problems and they are not fundamentally different from each other from the control point of view.

The usual approach is to classify systems to be controlled as either linear or non-linear. Consider for example a second-order system where

$$y''+ay'+by = u(t) \tag{5.1}$$

where y is the system response (dependent variable), u(t) is the input, which is a function of the independent variable (t), which is usually time. a and b are coefficients. It is assumed that the dependent variable y is a smooth function of t and is continuously differentiable up to any required order. y' and y'' represent the first and second differentials of y with respect to the independent variable, t.

The following cases arise:

i) If both a and b are constants, then the theory of the control of linear systems applies.
ii) It either a or b (or both) are functions of the independent variable (t), then the equation is still classified as linear, but relatively few solutions exist.
iii) If either a or b (or both) are functions of the dependent variable (y), then the system is classified as non-linear.

The standard linear control theory depends mainly upon the application of Laplace transforms. The general solution of equation 5.1 with constant coefficients comprises the sum of the complementary function, y_c, and the particular integral, y_p

$$y = y_c+y_p \tag{5.2}$$

This is the case of the theory of linear inhomogeneous differential equations from which it is known that if the system is disturbed from a quiescent state there will result a transient response that will either decay in time if there is positive damping, or grow without limit if there is negative damping. There will also be the forced response that will depend upon the nature of the applied input.

Another important factor is that limit-cycle type of self-excited vibrations are not possible in linear systems (but unstable self-excited vibrations are, as mentioned above). Limit-cycle self-excited vibrations are stable, finite amplitude oscillations; they are discussed later in Section 5.6.

In either linear or non-linear systems, the coefficients of the governing differential equation (if such exists) may have a finite number of discontinuities. In this case the governing equation and its solutions are both piecewise continuous.

The control of a process is conventionally effected by the use of one of the following types of controller, depending upon the nature of the application. These are well documented in many standard texts, for example, Watts and Cooper, (1999).

i) Proportional (P)
ii) Proportional plus differential (PD)
iii) Proportional plus integral (PI)
iv) Proportional plus integral plus differential (PID)

Usually, these controllers are modelled by linear differential equations with constant coefficients, and as mentioned earlier, theoretical solutions for the complete controller plus process are sought by the application of Laplace transforms. Even if the process is non-linear, the above types of controller may still be successfully applied in practice by tuning the controller using heuristic rules, which are described in standard texts.

The overall stability of linear systems with constant coefficients may be determined by the root-locus or the Routh-Hurwitz criteria. Minorsky, (1969), treats the stability theory of non-linear systems which have mathematical models by using Liapounov's criteria.

For systems that are capable of mathematical modelling and satisfactory solution, there is no need to consider FL control. But for other systems that do not have solutions which meet requirements FL control may offer the most effective option. Non-linearities in industrial systems are generally in one of two categories:

i) Those due to non-linear phenomena such as pressure-volumetric flow rate in turbulent pipe flow, or magnetic saturation. These are normally treated by linearising the relationships about the operating point.
ii) Dead-band or double valued phenomena such as backlash in mechanical systems, or hysteresis. These are normally treated by assuming that the output harmonics in periodic motion may be neglected as being of secondary importance and attenuated by the process system.

Such treatments may, or may not, meet requirements. There are also evolutionary systems which are better treated by FL control.

In the following treatment consideration is first given to linear systems, but to provide a foundation for subsequent developments involving FL non-linear systems this treatment is not based upon the conventional Laplace transforms method, rather, the mathematical models are recast in finite difference form, which requires responses to be evolved by iteration. The behaviour of various systems are examined and compared by applying a step input. The treatment can obviously be broadened to include any input pattern. Feedback single input-single output (SISO) type systems are considered here but feedforward can also be treated.

5.1 Second-order linear systems

Systems of any order may exist, but in practice those up to second-order are considered to be of particular importance and attention is now directed to these. Systems of higher order are not necessarily neglected, because even up to fourth-order their behaviour may be quite well approximated by second-order systems except for relatively small values of time.

Consider a single input, single output (SISO) system represented by

$$ay''+by'+cy = gu(t) \qquad 5.3)$$

where g,a,b and c are constants.

Now defining a characteristic time, $T=(a/c)^{1/2}$ and a characteristic linear dimension, $h=g/c$. Further, let the following dimensionless quantities be defined

$$z = y/h;\ s = t/T;\ k = b/(a/c)^{1/2}$$

Then equation 5.3) may be recast in dimensionless form as

$$z''+kz'+z = u(s) \qquad 5.4)$$

where u(s) is a dimensionless input and z = z(s).

The following cases arise:

Case	**k**	**Type**
I	>2	Overdamped
II	2	Critically damped
III	<2	Under damped
IV	0	Undamped

Now transforming equation 5.4) into a finite difference form using Taylor series expansions. A choice may be made between forward, central or backward finite difference forms of the time derivatives. For example, the central difference forms are

(First derivative) $z' = (z_{n+1}-z_{n-1})/2ds$ 5.5)

where ds is a small increment of s.

(Second derivative) $z'' = (z_{n+1}+z_{n-1}-2z_n)/ds^2$ 5.6)

Combining equations 5.4), 5.5) and 5.6) yields after some rearrangement

$$z_{n+1} = (u_n ds^2+z_n(2-ds)-z_{n-1}(1-kds/2))/(1+kds/2) \qquad 5.7)$$

With corresponding expressions for the forward and backward difference representations.

Assuming a step input

$$u(s) = H(s) \qquad 5.8)$$

where H(s) is Heaviside's step function: $H(0_-) = 0$; $H(0_+) = 1$

A choice is now made of a time interval such that, ds = 1. Also, for illustration, assume a case of sub-critical damping with k = 1. Then equation 5.7) degenerates to

$$z_{n+1} = 0.6667(1+z_n-0.5z_{n-1}) \qquad 5.9)$$

To develop z_n by iteration, it may be assume that (for example) $z_o = 0$ and $z_1 = 0.1$. The development of z over the first ten intervals; n = 1,2......10, is shown in Figure 5.1. This is a well-known result and verifies the backward difference treatment. The overdamped case; k = 3 is also shown. For this case

$$z_{n+1} = 0.4(1+z_n+0.5z_{n-1}) \qquad 5.10)$$

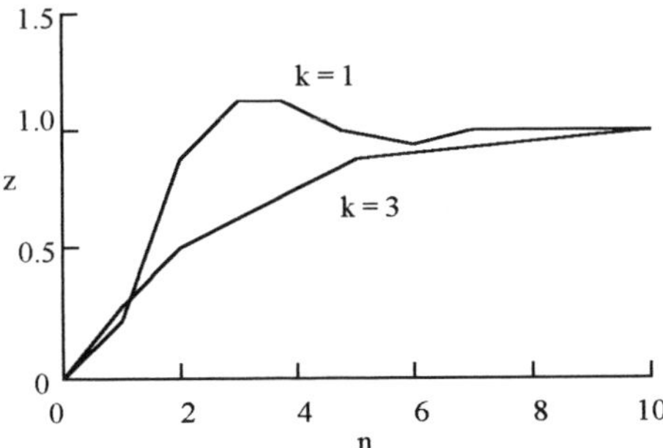

Figure 5.1 Iterative Response of a Linear Second-Order System to a Step-Input

5.2 Controllers

The classical types of controllers have been listed at the beginning. As examples of their behaviour, PD and PI controllers are considered below, each linked to a simple gain, k_o.

The PD controller is usually represented by a linear equation of the form

$$f_k u(t) = k_1 e + k_2 e' \qquad 5.11)$$

where e is the error, i.e. the difference between the reference input (desired value) and the output, e' = de/dt, also f_k, k_1 and k_2 are constants.

The controller, linked to a gain k_o in a feedback circuit is shown in Figure 5.2.

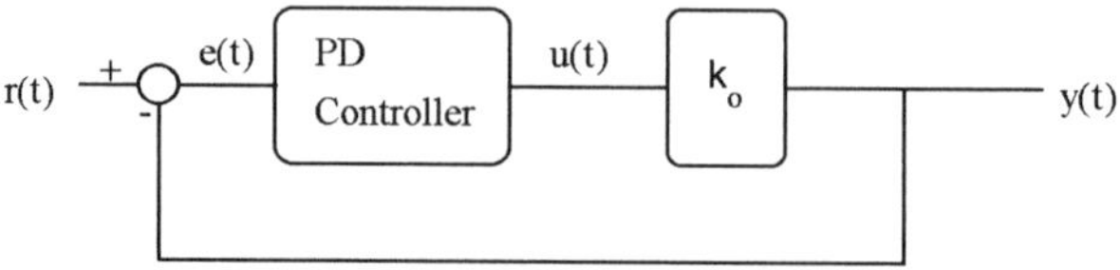

Figure 5.2 A PD Controller with a Process Gain k_o

In Figure 5.2 the controller output, u(t) is passed to the gain and emerges as the output, y(t), of the system. This is fed back to a differencing unit which compares it to the desired value, r(t). The difference is the error, e(t). The system process gain k_o is variable; $0<k_o<$infinity.

Now $$e(t) = r(t)-y(t) \qquad 5.12)$$

Hence, by combining equations 5.11) and 5.12), the following equation results

$$f_k u(t) = k_1(r(t)-y(t))+k_2(r'(t)-y'(t)) \qquad 5.13)$$

Defining the same scaling factors that were used to render equation 5.3) dimensionless, the following dimensionless form of equation 5.13) is found

$$u(s) = k_1`(r^+(s) -z(s))+k_2`(r^{+}{}'(s)-z'(s)) \qquad 5.14)$$

Now, $z(s) = k_o u(s)$. Hence equation 5.14) may be arranged to give

$$z'(s)+z(s)(k_1`+1/k_o)/k_2` = k_1`r^+(s)/k_2`+r^{+}{}'(s) \qquad 5.15)$$

In equations 5.14) and 5.15); $k_1` = k_1h/f_k$ and $k_2` = k_2h/(Tf_k)$ and $r^+ = r/h$.

It will suffice here to consider the case of $k_1`= k_2`=1$, which gives equal weight to the proportional and differential elements of the PD model, equation 5.11). Also for the purpose of comparison a step input is assumed, $r^+(s) = H(s)$. Equation 5.15) then degenerates to

$$z'+z(1+1/k_o)=1 \qquad 5.16)$$

where the bracket (s) is omitted. Performance specifications are often given in terms of the system response to a step change in input. Typical amongst these are the peak value of the output, percentage overshoot, 3 or 5% settling time and the rise time. They are treated in detail in most standard texts on automatic control and will not be further discussed here.

Now recasting equation 5.16) in finite difference form using a forward difference type of time derivative, the following expression is found

$$z_{n+1} = 1 - z_n/k_o \qquad 5.17)$$

It may be noted that a central difference form gives less satisfactory results in this case.

Assuming initial values; $z_o = 0$ and $z_1 = 0.2$ (for example), an iterative process enables successive z values to be generated. Figures 5.3a) and b) show the results for $k_o = 2$ and $k_o = 1/2$ respectively.

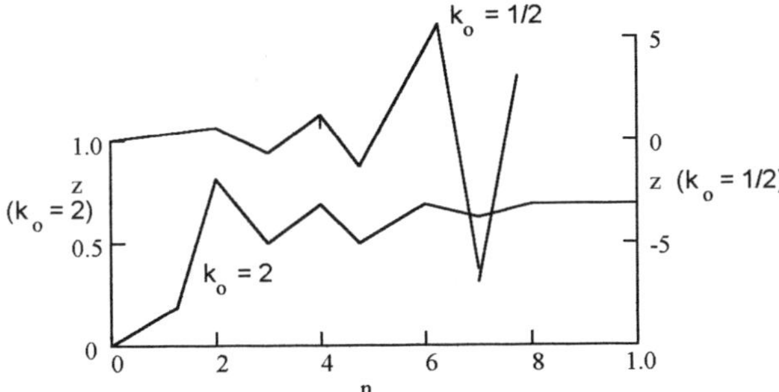

Figure 5.3 The Iterative Response of a PD Controller with a Process Gain $k_{o.}$

Considering next the PI controller, the equivalent model to equation 5.11) is

$$f_c u(t) = c_1\, e(t) + c_2 \quad e(t`)dt` \qquad 5.18)$$

where f_c,c_1 and c_2 are constants and t` is non-current time.

Differentiating equation 5.18) with respect to time, t, yields

$$f_c u'(t) = c_1 e'(t) + c_2 e(t) \qquad 5.19)$$

Comparing equation 5.19) with equation 5.11), it may be noted that they are of the same form, with u(t) replaced by u'(t). Therefore following a similar pattern to that used in obtaining equation 5.15), (noting that $z(s) = k_o u(s)$) the following expression is found

$$z(s) + z'(s)(c_1` + 1/k_o)/c_2` = c_1` r^+(s)/c_2` + r^{+}{}'(s) \qquad 5.20)$$

Now comparing this result with equation 5.15) (with $c_1` = c_2` = 1$)

$$z'(1 + 1/k_o) + z = 1 \qquad 5.21)$$

The forward difference form of this equation is

$$z_{n+1} = (1 + z_n/k_o)/(1 + 1/k_o) \qquad 5.22)$$

Assuming the same initial values as before; $z_o = 0$ and $z_1 = 0.2$, the iteration process generates z values for the PI controller as shown in Figure 5.4), for $k_o = 2$ and $k_o = 0.5$ These may be compared with there results previously obtained for the PD controller shown in Figures 5.3. The steady state values for the PD controller is <1. Whereas that for the PI controller is unity. This is a well-known result.

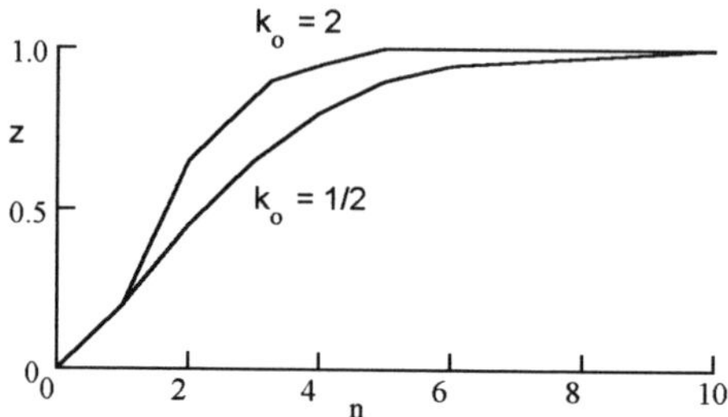

Figure 5.4 The Iterative Response of a PI Controller with a Process Gain k_o

A PID controller may be considered to be a linear combination of a PD and a PI controller. This implies that the error signal is split into two components

$$e(t) = e_k(t)+e_c(t) \qquad 5.23)$$

The result is two control outputs (u_k and u_c) which are added to produce the total output signal

$$u(t) = u_k(t)+u_c(t) \qquad 5.24)$$

A system which incorporates this principle is shown in Figure 5.5.

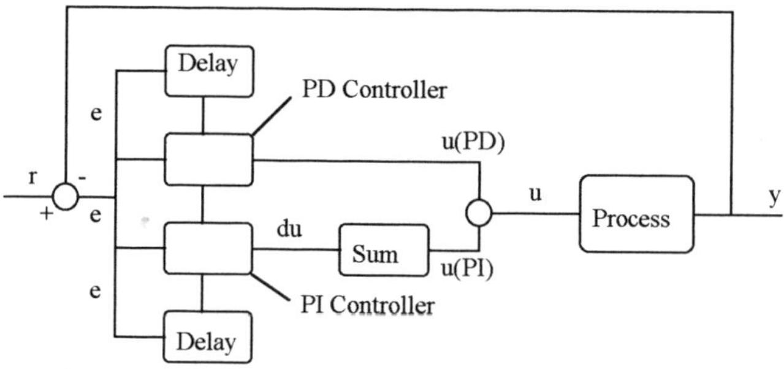

Figure 5.5 A PID Controller by PD and PI Controller Addition

It is unlikely in practice that a controller would be linked to a simple gain and attention is therefore now given to a second-order process. From equation 5.7), for an overdamped process with k = 3 and time increment ds = 1

$$z_{n+1} = 0.4(u_n+z_n+0.5z_{n-1}) \qquad 5.25)$$

From equation 5.14), with $k_1` = k_2` = 1$ and a step input applied, the PD controller gives

$$u = 1-z-z'$$

With a forward difference form of this equation the PD controller action is

$$u_n = 1-0.5(z_n+z_{n+1}) \quad 5.26)$$

Combining equations 5.25) and 5.26) yields

$$z_{n+1} = 0.3333(1+0.5(z_n+z_{n-1})) \quad 5.27)$$

An iterative process can again be used to generate z values starting from initial values; $z_o = 0$, $z_1 = 0.2$ (for example). The results of such a process based on equation 5.27) are shown in Figure 5.6. A step input has again been assumed, as before.

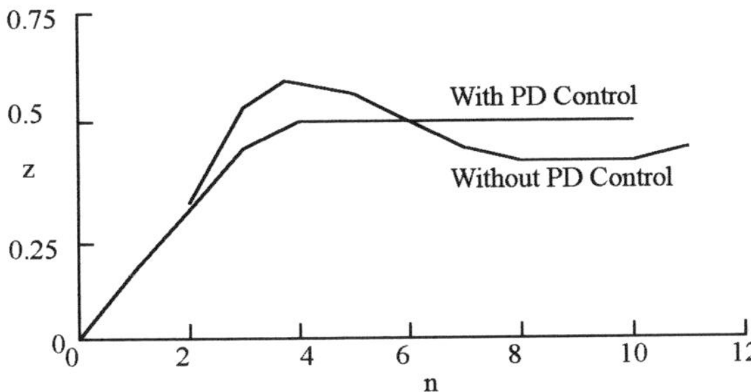

Figure 5.6 The Iterative Response of a PD Controller with a Second-Order Overdamped Linear Process

Consider now the action of a PI controller on the same overdamped second-order system (k = 3) as shown above. From equation 5.19) it may be deduced that the controller action is

$$u' = 1-z-z' \quad 5.28)$$

Recasting equation 5.28) in backward difference form

$$u_n = 1-2z_n+z_{n-1} \quad 5.29)$$

Choosing initial values of $z_o = 0$, $z_1 = 0.2$, then δu_1 in equation 5.29) is found. Substituting this in equation 5.25), the value of z_2 is found. Using this value and noting that $u_o = 0$, a second cycle of equation 5.29) provides δu_2. Now in equation 5.25) $u_2 = \delta u_1+\delta u_2$, hence z_3 may be found. Successive iterations enables higher z values to be found. Figure 5.7 illustrates the pattern, which is of the usual type with zero steady state error.

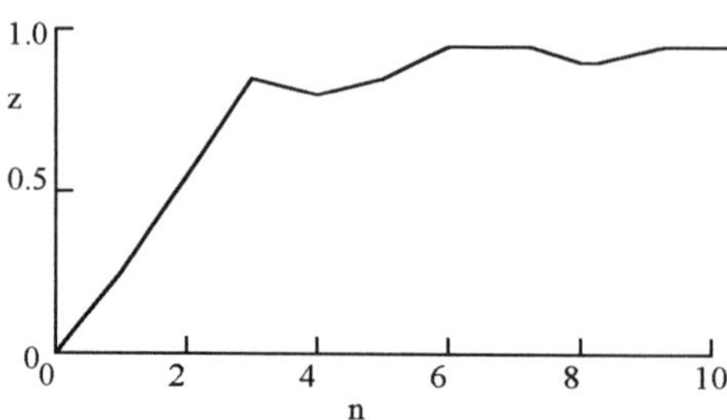

Figure 5.7 The Iterative Response of a PI Controller with a Second-Order Overdamped Linear Process

The effect on the response of an underdamped process by the addition of a PD controller is shown in Figures 5.8 for different degrees of underdamping, Namely, k = 0.5 and k = 1.0.

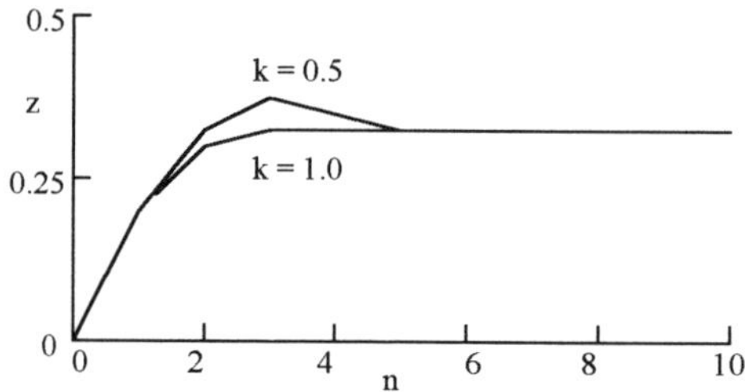

Figure 5.8 The Response of a PD Controller with Different Degrees of Underdamping

5.3 Non-linear systems

There are theoretical treatments of non-linear control systems, but they do not have the same generality as the linear theory. Use is made of linearisation and perturbation theory and there are a few analytical solutions, these treatments are described in several useful texts, such as that by Minorsky (1969). In practice, PID controllers can be successfully applied to non-linear systems , both for those with and those without mathematical models. Heuristic rules, such as the standard Ziegler-Nicolls methods, are used to tune the controllers under these circumstances. Fuzzy logic control (FLC) is also now finding increasingly wide application in such cases, all that is required of the system is that it shall be controllable, and that the objectives to be satisfied can be specified. FLC is therefore inherently less restrictive and of much wider scope than PID controllers, which can be emulated, if required.

5.4 Fuzzy logic control (FLC)

Clearly, there is no practical need to apply FLC to linear processes for which a PID controller can be specified, and which exist for a wide range of applications, if such will satisfy requirements. As an introduction to FLC however consideration is now given to a the FLC control of a second-order underdamped process that has previously been treated, see Figure 5.8.

For the controller the simplest type of membership class will be assumed, namely, singletons, that will be linked with simple partitioning of the universes of discourse which involve comparatively few logic relations between the antecedents and conclusions. The FLC will emulate a PD type control and the system will be subject to a step input, as in previously considered cases.

Normalised universes of discourse are considered, partitioned by singleton distributions as shown in Figures 5.9a),b) and c).

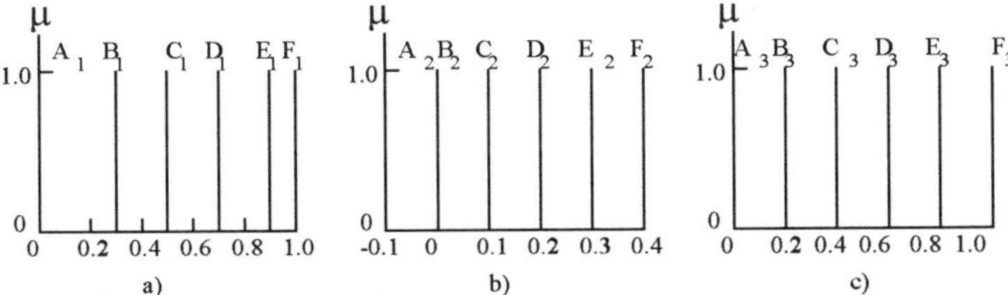

Figure 5.9 Partitioning of the Normalised Universes of Discourse
a) z b) δz c) u

Table 5.1 is a rule-base form which will emulate a PD controller. z and δz are the antecedents of the FL proposition

IF Z AND δZ THEN U

Table 5.1. The Rule-Base for u

z \ **δz**		A_2	B_2	C_2	D_2	E_2	F_2
A_1	A_3	A_3	B_3	C_3	C_3	D_3	
B_1	A_3	B_3	C_3	C_3	C_3	D_3	
C_1	B_3	C_3	C_3	C_3	D_3	E_3	
D_1	C_3	C_3	D_3	D_3	E_3	E_3	
E_1	C_3	D_3	D_3	E_3	E_3	E_3	
F_1	D_3	D_3	E_3	E_3	E_3	E_3	

The process considered is a second-order type given in general terms by equation 5.7). Assuming a unit time increment (ds = 1) and an underdamped system with k = 1, then equation 5.7) degenerates to

$$z_{n+1} = 0.6667(u_n+z_n-0.5z_{n-1}) \quad 5.30)$$

Which is the more general form of equation 5.9).

To start the iteration process, initial values are assumed; $z_o = 0$, $z_1 = 0.2$ and $\delta z_1 = 0.2$. From Figure 5.9 the relevant membership values of z_1 and δz_1 are obtained, and from Table 5.1 the linguistic sets of u are obtained, thus

		IF Z	AND δZ	THEN U
$z_1 = 0.2$	$\delta z_1 = 0.2$			
$\mu_{B1} = 0.667$	$\mu_{D2} = 1.0$	B_1	D_2	C_3
$\mu_{A1} = 0.333$		A_1	D_2	C_3

Defuzzifying the u linguistic terms

$$u_1 = (C_3\min(0.667,1.0)+C_3\min(0.333,1.0))/\Sigma\mu_i$$
$$= 0.6(0.667+0.333)/(0.667+0.333) = 0.6$$

From equation 5.30)

$$z_2 = 0.6667(u_1+z_1-0.5) = 0.667(0.6+0.2-0) = 0.533$$

and $\delta z_2 = z_2 - z_1 = 0.3333$

This completes the first cycle. The second cycle may be started by finding the membership values for z_2 and δz_2 using again the information in Figure 5.9). The results are as follows:

		IF Z	AND δZ	THEN U
$z_2 = 0.533$	$\delta z_2 = 0.333$			
$\mu_{D1} = 0.165$	$\mu_{F2} = 0.33$	D_1	F_2	E_3
$\mu_{C1} = 0.835$	$\mu_{E2} = 0.67$	D_1	E_2	E_3
		C_1	F_2	E_3
		C_1	E_2	D_3

Defuzzifying the u linguistic terms

$$u_2 = (0.2\min(0.165,0.33)+0.2\min(0.165,0.67)+0.2\min(0.835,0.33)+ 0.4\min(0.835,0.67))/(0.165+0.835+0.33+0.67)$$

Hence, $u_2 = 0.301$

From equation 5.30)

$$z_3 = 0.667(u_2+z_2-0.5z_1)$$
$$= 0.667(0.301+0.533-0.5*0.2)$$

Hence, $z_3 = 0.489$ and $z_3 = z_3 - z_2 = -0.044$

Continuing with the same method, the following results are obtained:

		IF Z	AND δZ	THEN U	
$z_3 = 0.489$	$\delta z_3 = -0.044$	C_1	A_2	B_3	$u_3 = 0.709$
$\mu_{C1} = 0.945$	$\mu_{A2} = 0.44$	C_1	B_2	C_3	$z_4 = 0.621$
$\mu_{B1} = 0.055$	$\mu_{B2} = 0.56$	B_1	A_2	A_3	$\delta z_4 = 0.132$
		B_1	B_2	B_3	
$z_4 = 0.621$	$\delta z_4 = 0.132$	D_1	D_2	D_3	$u_4 = 0.487$
$\mu_{D1} = 0.605$	$\mu_{D2} = 0.320$	D_1	C_2	D_3	$z_5 = 0.576$
$\mu_{C1} = 0.335$	$\mu_{C2} = 0.680$	C_1	D_2	C_3	$\delta z_5 = -0.045$
		C_1	C_2	C_3	
$z_5 = 0.576$	$\delta z_5 = -0.45$	D_1	A_2	C_3	$u_5 = 0.651$
$\mu_{D1} = 0.380$	$\mu_{A2} = 0.451$	D_1	B_2	C_3	$z_6 = 0.611$
$\mu_{C1} =.0.620$	$\mu_{B2} = 0.549$	C_1	A_2	B_3	$\delta z_6 = 0.035$
		C_1	B_2	C_3	
$z_6 = 0.611$	$\delta z_6 = 0.035$	D_1	C_2	D_3	$u_6 = 0.559$
$\mu_{D1} = 0.555$	$\mu_{C2} = 0.352$	D_1	B_2	C_3	$\delta z7 = 0.588$
$\mu_{C1} = 0.445$	$\mu_{B2} = 0.648$	C_1	C_2	C_3	$\delta z_7 = -0.023$
		C_1	B_2	C_3	
$z_7 = 0.588$	$\delta z_7 = -0.023$	D1	A_2	C_3	$u_7 = 0.677$
$\mu_{D1} = 0.440$	$\mu_{A2} = 0.231$	D_1	B_2	C_3	$z_8 = 0.639$
$\mu_{C1} = 0.560$	$\mu_{B2} = 0.769$	C_1	A_2	B_3	$\delta z_8 = 0.051$
		C_1	B_2	C_3	
$z_8 = 0.639$	$\delta z_8 = 0.051$	D_1	C_2	D_3	$u_8 = 0.536$
$\mu_{D1} = 0.695$	$\mu_{C2} = 0.514$	D_1	B_2	C_3	$z_9 = 0.588$
$\mu_{C1} = 0.305$	$\mu_{B2} = 0.486$	C_1	C_2	C_3	$\delta z_9 = -0.052$
		C_1	B_2	C_3	
$z_9 = 0.588$	$\delta z_9 = -0.052$	D_1	A_2	C_3	$u_9 = 0.655$
$\mu_{D1} = 0.438$	$\mu_{A2} = 0.517$	D_1	B_2	C_3	$z_{10} = 0.615$
$\mu_{C1} = 0.562$	$\mu_{B2} = 0.483$	C_1	A_2	B_3	$\delta z_{10} = 0.0278$
		C_1	B_2	C_3	

The successive values of z_n are shown in Figure 5.10, which may be compared with Figure 5.8.

It is, of course, possible to tune an initial FLC scheme such as the one shown above by adjustment of the partitioning of the universes of discourse and to the rule-base. The approach to, and final steady state value can be modified in this way.

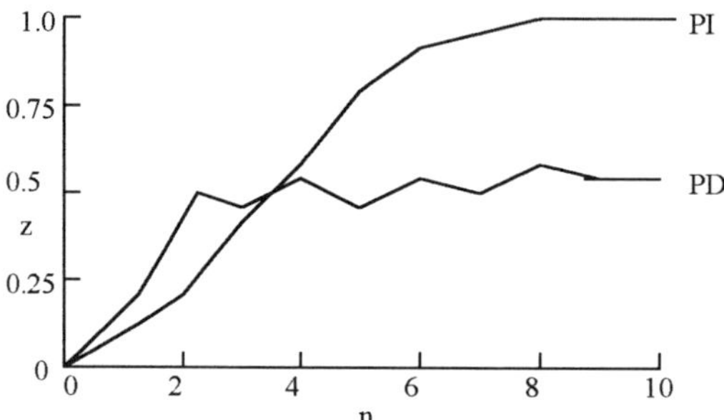

Figure 5.10 FLC-PD and PI Control of a Second-Order Linear Underdamped Process

The corresponding FLC-PI emulation is given below. Instead of equation 5.30), the following process equation for a linear second-order process with a unit step input is appropriate

$$z_{n+1} = 0.6667(\Sigma\delta u_n + z_n - 0.5z_{n-1}) \qquad 5.31)$$

The same form of rule-base is used as for the FLC-PD emulation, but the controller output action universe of discourse is now in terms of the difference values as shown in Figure 5.11.

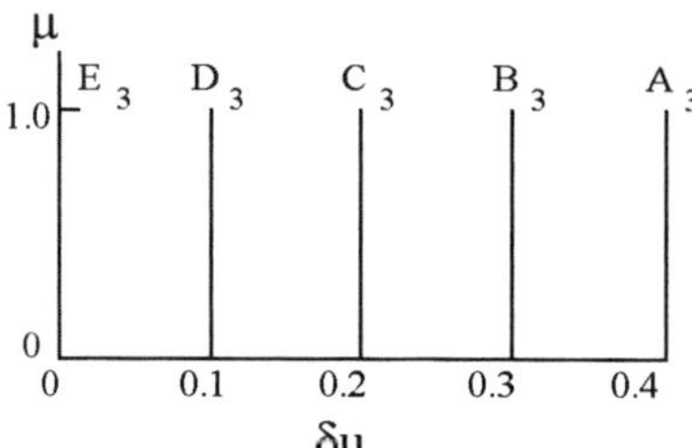

Figure 5.11 Partitioning of the δu Universe of Discourse

Starting the iteration process again with initial values; $z_0 = 0$, $z_1 = 0.2$ and $\delta z_1 = 0.2$. The same universes of discourse and partitioning are used as for the above FLC-PD case. Therefore the rule-base, Table 5.1, and defuzzification using Figure 5.11 gives a proposition of the form

IF Z	AND δZ	THEN δU	
B_1	C_2	C_3	$\delta u_1 = 0.2$
A_1	C_2	C_3	$z_2 = 0.667(0.2+0.2-0)$
			$= 0.267$
			and $\delta z_2 = 0.067$

Second and subsequent cycles provide the following development of z:

		IF Z	AND δZ	THEN δU	
					$\delta u_2 = 0.255$
$z_2 = 0.267$	$\delta z_2 = 0.067$	B_1	C_2	C_3	$\Sigma\delta u_n = \delta u_1+\delta u_2$
$\mu_{B1} = 0.899$	$\mu_{C2} = 0.666$	B_1	B_2	B_3	$= 0.455$
$\mu_{A1} = 0.111$	$\mu_{B2} = 0.334$	A_1	C_2	B_3	$z_3 = 0.414$
		A_1	B_2	A_3	$\delta z_3 = 0.148$
$z_3 = 0.414$	$\delta z_3 = 0.148$	C_1	D_2	C_3	$\delta u_3 = 0.2$
$\mu_{C1} = 0.570$	$\mu_{D2} = 0.480$	C_1	C_2	C_3	$\Sigma\delta u_n = 0.655$
$\mu_{B1} = 0.430$	$\mu_{C2} = 0.520$	B_1	D_2	C_3	$z_4 = 0.624$
		B_1	C_2	C_3	$\delta z_4 = 0.209$
$z_4 = 0.624$	$\delta z_4 = 0.209$	D_1	E_2	E_3	$\delta u_4 = 0.125$
$\mu_{D1} = 0.620$	$\mu_{E2} = 0.910$	D_1	D_2	D_3	$\Sigma\delta u_n = 0.779$
$\mu_{C1} = 0.380$	$\mu_{D2} = 0.910$	C_1	E_2	D_3	$z_5 = 0.797$
		C_1	D_2	C_3	$\delta z_5 = 0.174$
$z_5 = 0.797$	$\delta z_5 = 0.174$	E_1	D_2	E_3	$\delta u_5 = 0.068$
$\mu_{E1} = 0.495$	$\mu_{D2} = 0.740$	E_1	C_2	D_3	$\Sigma\delta u_n = 0.847$
$\mu_{D1} = 0.515$	$\mu_{C2} = 0.260$	D_1	D_2	D_3	$z_6 = 0.888$
		D_1	C_2	D_3	$\delta z_6 = 0.091$
$z_6 = 0.888$	$\delta z_6 = 0.091$	E_1	C_2	D_3	$\delta u_6 = 0.105$
$\mu_{E1} = 0.940$	$\mu_{C2} = 0.913$	E_1	B_2	D_3	$\Sigma\delta u_n = 0.952$
$\mu_{D1} = 0.060$	$\mu_{B2} = 0.087$	D_1	C_2	D_3	$z_7 = 0.962$
		D_1	B_2	C_3	$\delta z_7 = 0.073$
$z_7 = 0.962$	$\delta z_7 = 0.073$	F_1	C_2	E_3	$\delta u_7 = 0.060$
$\mu_{F1} = 0.620$	$\mu_{C2} = 0.733$	F_1	B_2	D_3	$\Sigma\delta u_n = 1.012$
$\mu_{E1} = 0.380$	$\mu_{B2} = 0.267$	E_1	C_2	D_3	$z_8 = 1.020$
		E_1	B_2	D_3	$\delta z_8 = 0.058$
$z_8 = 1.020$	$\delta z_8 = 0.058$	F_1	C_2	E_3	$\delta u_8 = 0.042$
$\mu_{F1} = 1.000$	$\mu_{C2} = 0.582$	F_1	B_2	D_3	$\Sigma\delta u_n = 1.054$
	$\mu_{B2} = 0.418$				$z_9 = 1.062$
					$\delta z_9 = 0.042$

$z_9 = 1.062$	$\delta z_9 = 0.042$	F_1	C_2	E_3	$\delta u_9 = 0.058$
$\mu_{F1} = 1.000$	$\mu_{C2} = 0.421$	F_1	B_2	D_3	$\Sigma\delta u_n = 1.112$
	$\mu_{B2} = 0.579$				$z_{10} = 1.109$
					$\delta z_{10} = 0.047$

The z_n curve is also shown in Figure 5.10 and is of the normal PI form. This control action clearly has a significant moderating influence on the process.

5.5 Second-order non-linear processes

The previous type of FLC can be extended to non-linear processes. Consider for example, a process with a non-linear damping term, z^b, as in equation 5.32

$$z'' + az^b z' + z = u(t) \qquad 5.32)$$

Where a and b are constants.

Recasting this model into finite-difference form using central difference forms for both time derivatives, the following expression is obtained

$$z_{n+1} = (u_n + z_n - (1 - az_n^b)z_{n-1})/(1 + az_n^b) \qquad 5.33)$$

In the following example, values of a=1 and b=0.6 will be assumed for illustration purposes. Also singleton FL sets will be used and the universes of discourse are partitioned as shown in Figure 5.12.

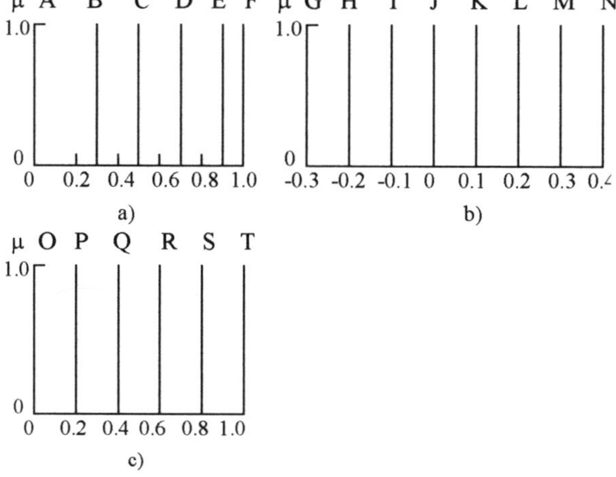

Figure 5.12 Partitioning of the Universes of Discourse
a) z b) δz c) u

Table 5.2 A Rule-Base for PD Emulation

	δz							
z	G	H	I	J	K	L	M	N
A	O	O	O	O	P	Q	Q	Q
B	O	O	O	P	Q	Q	Q	R
C	O	O	P	Q	Q	Q	R	S
D	O	P	Q	Q	R	R	S	S
E	P	Q	Q	R	R	S	S	S
F	Q	Q	R	S	S	S	S	S

For the FLC-PD emulation the iteration is started with assumed initial values; $z_0 = 0$, $z_1 = 0.2$ and $\delta z_1 = 0.2$. Following the same scheme as before, the first two cycles proceed as follows:

$z_1 = 0.200$	$\delta z_1 = 0.200$	IF Z	AND δZ	THEN δU	$u_1 = 0.600$
$\mu_{B1} = 0.333$	$\mu_{C2} = 1.000$	B	I	Q	$z_2 = 0.588$
$\mu_{A1} = 0.667$		A	I	Q	$\delta z_2 = 0.488$

The updated values; u ,z and δz, are then the inputs for the second iteration yielding

$z_2 = 0.588$	$\delta z_2 = 0.388$	D	K	S	$u_2 = 0.219$
$\mu_{D1} = 0.440$	$\mu_{E2} = 0.880$	D	J	S	$z_3 = 0.499$
$\mu_{C1} = 0.560$	$\mu_{D2} = 0.120$	C	K	S	$\delta z_3 = -0.090$
		C	J	R	

After nine cycles the output (z) levels off to a steady state value of about 0.3. The following table shows the evolution of the output.

Table 5.3 Iterative Values of Equation 5.33)

Cycle	$\mathbf{z_n}$	$\boldsymbol{\delta z_n}$	$\mathbf{u_n}$	$\mathbf{z_{n+1}}$
1	0.200	0.200	0.600	0.588
2	0.588	0.388	0.219	0.499
3	0.499	-0.090	0.290	0.296
4	0.296	-0.202	0.400	0.250
5	0.256	-0.040	0.354	0.310
6	0.310	0.054	0.237	0.285
7	0.285	-0.026	0.333	0.308
8	0.308	0.023	0.271	0.292
9	0.292	-0.015		

The z_n values for the above FLC-PD case are shown in Figure 5.13). These values may be compared with the z_n values for the output of the process with a step input; $u(t) = H(t)$ and no control, that is, for the process model given by equation 5.34)

$$z'' + z^{0.6}z' + z = 1 \qquad 5.34)$$

The process output for this no-control response is also shown in Figure 5.13).It may be seen that the main effect of the FLC-PD action is that of attenuation of the output.

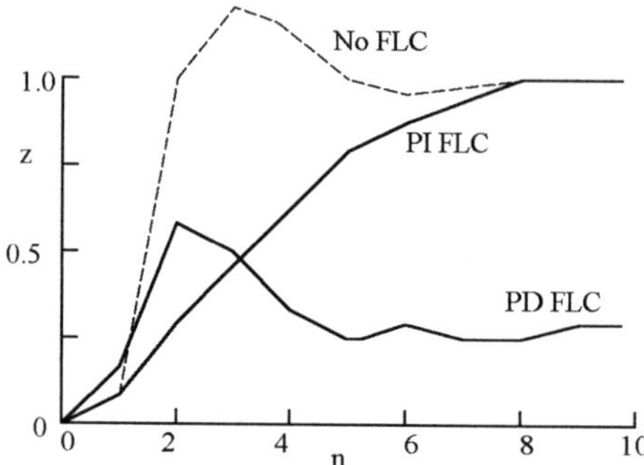

Figure 5.13 FLC-PD and PI Control of a Second-Order non-Linear Process

For comparison, consider now the same process, but with an associated FLC-PI controller. Assuming the same values of a and b in equation 5.33), namely 1 and 0.6 respectively, the equivalent form of central difference equation is

$$z_{n+1} = (\Sigma\delta u_i + z_n - (1 - z_n^{0.6}))/(1 + z_n^{0.6}) \qquad (i = 1 \text{ to } n) \qquad 5.35)$$

The same form of rule-base is used as for the FLC-PD emulation (Table 5.3), also the same z and δz partitioning as in Figure 5.12. (The δu partitioning is as shown in Figure 5.11).

For PI control, the evolution of z for nine cycles is given in Table 5.4 below, and the results are also displayed in Figure 5.13). Comparing this output with that of the process with no control action, it may be noted that the asymptotic values are both about unity, but that there is virtually complete smoothing of the output so that it rises monotonically towards the steady state value.

Table 5.4 Iterative Values of Equation 5.35)

Cycle	z_n	δz_n	u_i	z_{n+1}
1	0.200	0.200	0.190	0.258
2	0.258	0.058	0.401	0.476
3	0.476	0.218	0.587	0.672
4	0.672	0.196	0.705	0.784
5	0.784	0.112	0.843	0.869
6	0.869	0.085	0.954	0.959
7	0.959	0.090	1.002	1.018
8	1.018	0.0560	1.002	1.027
9	1.027			

Common non-linear damping terms arise from Coulomb (dry) friction, (sign)f and also from turbulent damping, z'^b, where for ideal fully turbulent conditions b=2, but more generally $1.5<b<2$. These would provide second-order models of the following form

$$z''+(\text{sign}z)af+z = u(t) \qquad \text{(Coulomb friction)} \qquad 5.36)$$

where f is a dimensionless force and a is a scaling factor. The equivalent finite difference form is

$$z_{n+1} = u_n+2z_n-z_{n-1}-(\text{sign}z)af \qquad 5.37)$$

And for turbulent damping

$$z''+cz'^b+z = u(t) \qquad 5.38)$$

Equivalently

$$z_{n+1} = u_n+z_n-c(\delta z_n)^b-z_{n-1} \qquad 5.39)$$

Where the central difference is used for z'' and the backward difference for z'. Equations 5.37) and 5.39) are the FLC-PD forms. In the FLC-PI form the u_n terms are replaced by $\Sigma\delta u_i$

Examples of non-linear dynamic equations may also be found in electronic and electrical engineering. A typical case would be that of the well-known van der Pol equation

$$z''-\varepsilon(1-z^2)z'+z = u(t) \qquad 5.40)$$

This finds applications both in the fields of radio communications and also in d.c. motor/generator pairs for example. Standard texts treat the so-called relaxation oscillations obtained when u(t) = 0 in equations 5.37) and 5.39). Non-linear saturation effects are infact found in many electrical and electronic phenomena.

The non-linear damping term in the van der Pol equation, $-\varepsilon(1-z^2)z'$, comprises a -ve linear component, $-\varepsilon z'$, which will produce self-excited oscillations when the process is perturbed from its quiescent state. There is also a +ve non-linear component, $\varepsilon z^2 z'$, which is overshadowed by the linear component for small amplitude oscillations. The amplitude will thus initially increase in value until the energy input rate by the linear component is balanced by the energy dissipation rate of the non-linear component. This provides the phenomenon of limit cycles; stable bounded self-excited oscillations which are only possible in non-linear processes. (Self-excited oscillations are unbounded in linear processes).

Some non-linear effects may be reduced to piecewise linear sectors. Such is the case with a mass moving with clearance between linear springs (known as backlash) and also when a mass is constrained by preloaded linear springs. Non-linear phenomena in several contexts are treated in other chapters in this text.

Another frequently quoted non-linear process governing equation has the form

$$z'' + cz' + (az + bz^3) = u(t) \qquad (a,b \text{ and } c>0) \qquad 5.41)$$

For the case of u(t) = 0, it known as the Duffing equation. This and the other process equations quoted above may all be the subjects of intervention by FLC, as discussed previously.

In some applications, the non-linear terms in the process model may only be available in graphical or tabular form. Such data may be easily and directly incorporated into a FLC scheme. Stochastic processes are of course possible, where for a given controller action the output of the process is probabilistic, this type too may be the subject of a FLC scheme; such cases are outside the scope of this text.

In the above discussion cases have been considered in which processes may be modelled sufficiently well by mathematical expressions. In more complex cases where mathematical models are unavailable the process may also be treated by a FL strategy, so that the whole system of process plus controller is FL based.

5.6 More general parameters

Up to this point, attention has been restricted to systems which are subject to a step change of reference signal, which in practice yields useful system performance data and enables comparisons to be made of behavioural differences. It will be appreciated that in the general case +ve and -ve values of the system variables must be admitted.

Examples of more general universes of discourse are shown below in Figure 5.14.

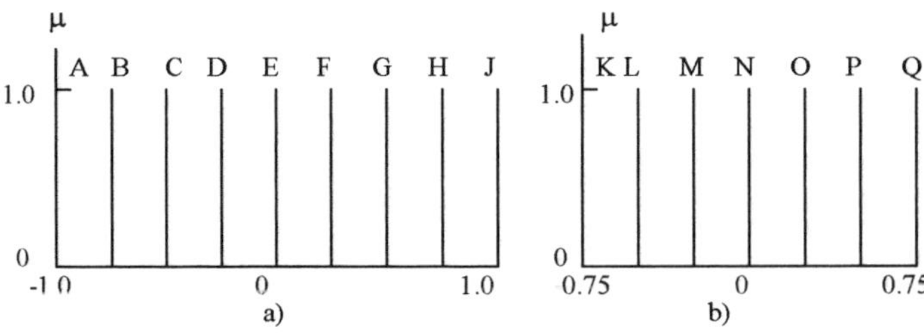

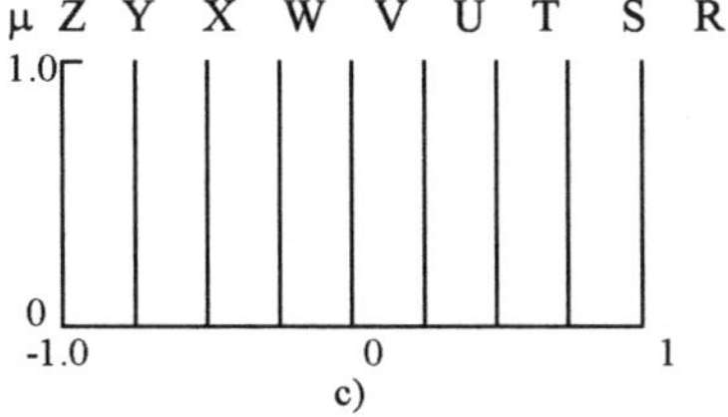

Figure 5.14 Typical FLC Universes of Discourse
a) z_n b) δz_n c) u_n

The rule-base would also need to be extended in this case and an example is given below in Table 5.5.

Table 5.5 Typical General Form of Rule-Base

		δz					
z	K	L	M	N	O	P	Q
A	R	R	R	S	T	T	U
B	R	R	S	T	T	U	U
C	R	S	T	T	U	U	V
D	S	T	T	U	U	V	W
E	T	T	U	U	V	W	W
F	T	U	U	V	V	W	X
G	U	U	V	W	W	X	Y

As usual, the singleton form of the fuzzy sets in Figure 5.14 may be recast in other more suitable shapes as circumstances demand. A fully articulated form of the process model (if such exists) is also required. For example, equation 5.32) would be recast in the form

$$z'' + (\text{sign}) a z^b z' + z = u(t) \qquad 5.42)$$

5.7 Look-up tables

The numerical results of the iterative process as described above enables a tabular form to be established, called a look-up table. This permits direct entry of the inputs and, by interpolation, the associated system response is obtained. For example an iterative process performed using the set distributions in Figure 5.14 and the rule-base in Table 5.5 enables a look-up table such as that shown in Table 5.6 to be formulated.

Table 5.6 Example of a FLC-PD Look-Up Table

		δz						
z	-0.3	-0.2	-0.1	0	0.1	0.2	0.3	0.4
0	1.0	1.0	1.0	1.0	0.8	0.6	0.6	0.6
0.3	1.0	1.0	1.0	0.8	0.6	0.6	0.6	0.4
0.5	1.0	1.0	0.8	0.6	0.6	0.6	0.4	0.2
0.7	1.0	0.8	0.6	0.6	0.4	0.4	0.2	0.2
0.9	0.8	0.6	0.6	0.4	0.4	0.2	0.2	0.2
1.0	0.6	0.6	0.4	0.2	0.2	0.2	0.2	0

An interesting feature of the tabular form of presentation of the numerical results is that the trajectory of the evolution of the system response, z,, can be portrayed in the (z, δz) space and this corresponds with the state-space (sometimes called the phase-space or phase-plane) representation, usually found in non-linear dynamics texts. Typical FLC-PD and PI trajectories commencing from (0,0) for a stable system are shown in Figure 5.15, assuming a unit step change of the reference input. Points such as A and B in Figure 5.15 are called singularities; A is a focus and B is a node. For an unstable system, which is originally in a quiescent state, the direction is reversed and the system unwinds from its original state if it is perturbed, as shown in Figure 5.16 for a unit step change. Another type of singularity is called a saddle point, found with an inverted pendulum system.

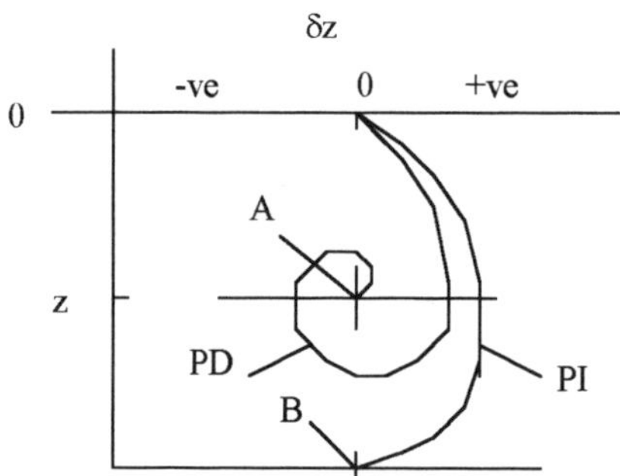

Figure 5.15 Typical FLC-PD and -PI Trajectories for Stable Systems with a Step Input

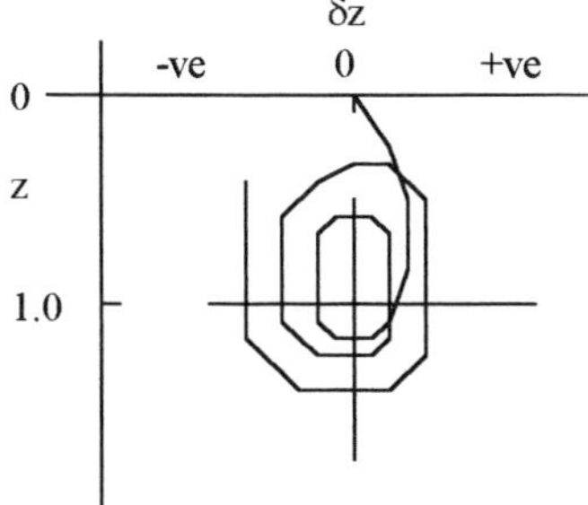

Figure 5.16 A Trajectory for an Unstable System

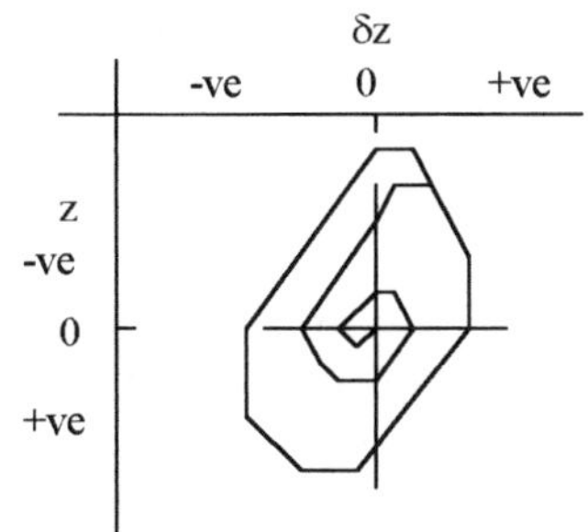

Figure 5.17 A Trajectory for a Stable Limit Cycle

Oscillations are generally undesirable in a control system and therefore a design objective would be to reduce the number of cycle loops as far as possible within the system performance specification.

Figure 5.17 shows a stable limit cycle type of trajectory commencing at (0,0) which may be obtained with a non-linear process. The system track winds itself onto a fixed cycle from any other non-closed cycle, a representative point would remain indefinitely on the cycle. Conversely an unstable system track would unwind itself from the limit cycle. In more complicated cases there may be concentric stable and unstable limit cycles and also more than one singular point in the plane.

Relating trajectories to the appropriate look-up table is only feasible if partitioning of the universes of discourse is sufficiently fine for adequate definition of the curve. For example, a 3x3 array in the table would be too coarsely structured to permit portrayal of any trajectory. Conversely, the finer the partitioning, then clearly the better the trajectory definition. It is known that the fundamentals of the study of oscillations and the quiescent state in non-linear control theory are similar to non-linear dynamics, though the former is more complicated due to the greater dimensionality. In particular, FLC characteristics are less easy to identify. The stability of FLC non-linear systems also has no formal basis at present. Each case is examined individually.

5.8 Hierarchical and adaptive FLC

The control structures discussed previously can be developed in several ways to achieve more fluent control of a process. Two prominent directions of development are:

i) Hierarchical control
ii) Adaptive control

There are many possible specific combinations of these two types, but in both cases there is a supervisory element to manage the behaviour of the system. This supervisory element interrogates the current FLC response and from its knowledge of what is required, it selects and implements a new controller mode from the available patterns. The two types of control are outlined in the following sections.

5.9 Hierarchical or distributed FLC

In this type of system the control activity is distributed amongst several units which have specialised functions within the overall control organisation.. For example, if there is need for precise control over a wide range of output variables, then this might be achieved by an actuator with coarse and fine adjustments, each with its own FLC. The supervisory element would select when to change the control initiative from the coarse (primary control) to the fine (secondary) control. The change-over decision may be made using the magnitude of the error signal as a metric. Figure 5.18 illustrates the elements of such a control system.

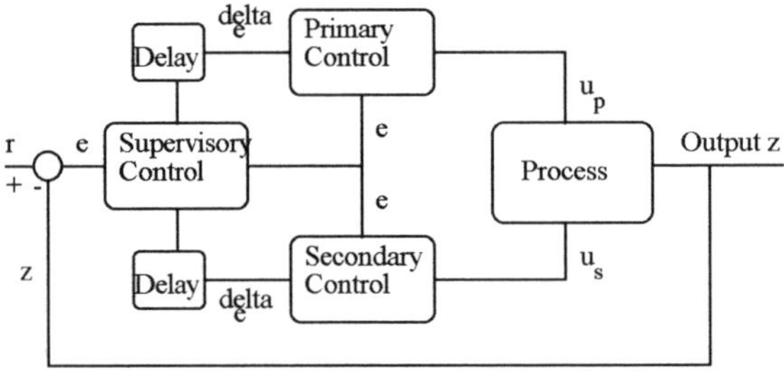

Figure 5.18 The Elements of a Hierarchical Control System

In the phase-space the trajectories for the primary and secondary controls would be as shown on Figures 5.19 and 5.20, for example. In the first figure for primary control the trajectory commences at the point representing the initial conditions and terminates on the boundary of the zone of secondary control. The continuation of the trajectory is shown in Figure 5.20. where it terminates on the point representing the final control output.

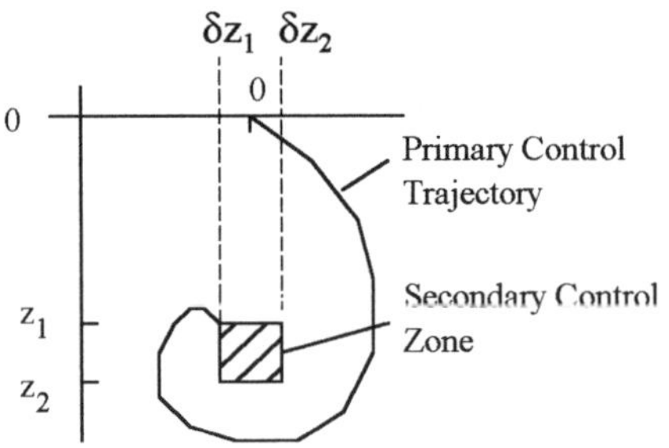

Figure 5.19 Primary Control Trajectory in the z, δz Space

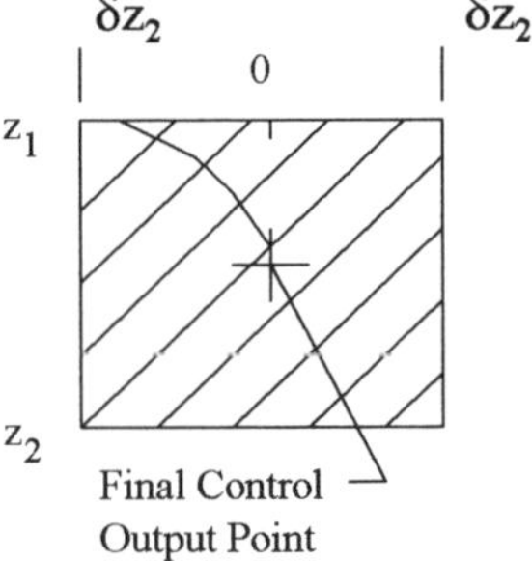

Figure 5.20. Secondary Control Trajectory in the z, δz Space

It is also possible to have mixed fuzzy and conventional control whereby, for example, the primary control is fuzzy logic based whilst the secondary control is a conventional type of PID controller.

5.10 Autoadaptive FLC

Up to this point the controllers discussed respond to set point changes or load disturbances in an invariable way. They are tuned either on-line or off-line through human intervention. The controllers themselves have a set pattern of response even though in a distributed system the control function may be switched between different controllers; the logic architecture is static.

In an autoadaptive controller the objects in the logic space may change in response to process or operating environment changes, or both. This self-reorganisation causes autonomous variations in the process controller input/output relations. Principal problems that present themselves for which autoadaptive FLC is appropriate are

i) Process or operating environment drift from the initial conditions.
ii) Transients due to changes in the set point or the load, or both.

There are several ways of resolving problems in the above categories, for example:

i) Use of a gain (or scaling) factor for the inputs or outputs.
ii) Rule-weights for factoring the logic propositions. Also known as truth factors.
iii) Adjustments of the linguistic sets supports. (For triangular sets, this means varying the width of the triangular base).
iv) Hedging of the linguistic sets. This means intensifying or diluting the linguistic sets by raising set membership values to some power which is >1 for intensifying and <1 for dilution.

In all cases a supervisory control element effects the logic adaptations by interrogating the process controller, applying logic analysis to deduce the level of changes required and then implementing the required changes.

As an example, consider a process in which a chemical reaction takes place as a fluid stream passes through a reactor containing a catalyst, as shown in Figure 5.21. Over a period of time the catalytic bed becomes contaminated and the reactants therefore require a longer residence time to achieve the same degree of reaction if the temperature is constant. This means reducing the liquid reactant flow rates. The controller therefore monitors the degree of reaction and adjusts the flow rates accordingly. The controller's temperature set point remains fixed.

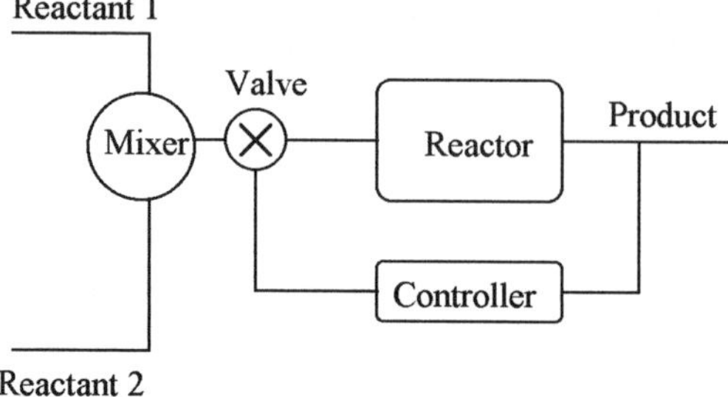

Figure 5.21 Control of a Chemical Process

Let z be the volumetric flow rate of the mixed reactants. The initial controller operating point on the e, δe space will be at point P for example. After some time it has moved to point Q, as shown in Figure 5.22.

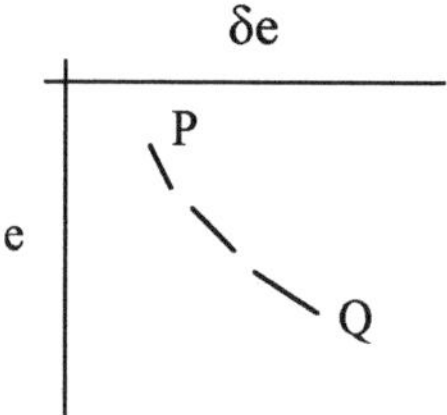

Figure 5.22 Controller Operating Point Drift on the e-δe Plane

This is a case where rule-weighting could be considered to return the controller operating point back to P, whilst the process operates at point Q.

Figure 5.23 represents a simple autoadjusting FLC for implementing a rule-weight management control supervisor. The input error (e) and the output (u) of the FL process controller are sampled by the supervisory FLC. The sample values are fuzzified and membership values of the fuzzy sets are evaluated using fuzzy set diagrams as shown in Figure 5.24. Figure 5.25 shows a fuzzy diagram for the supervisory output (v) sets.

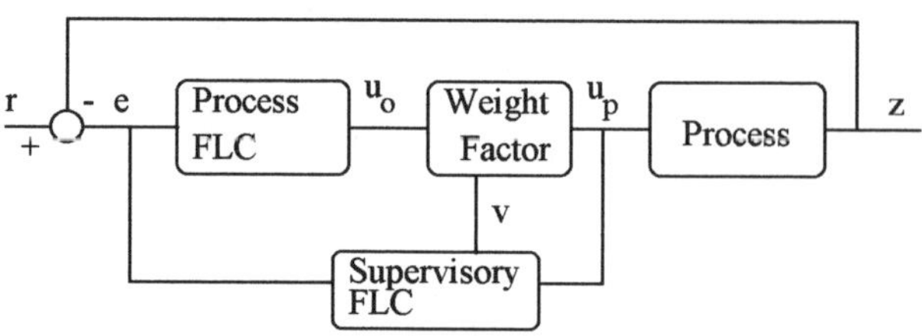

Figure 5.23 A Simple Weight Adjusting Self-Organising FLC

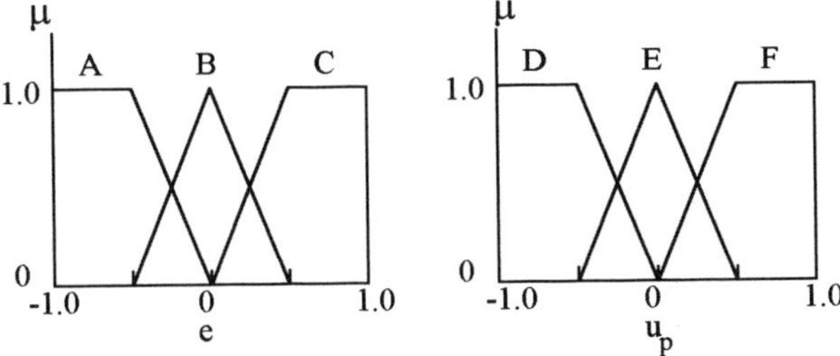

Figure 5.24 Normalised Fuzzy Diagrams for e and u_p

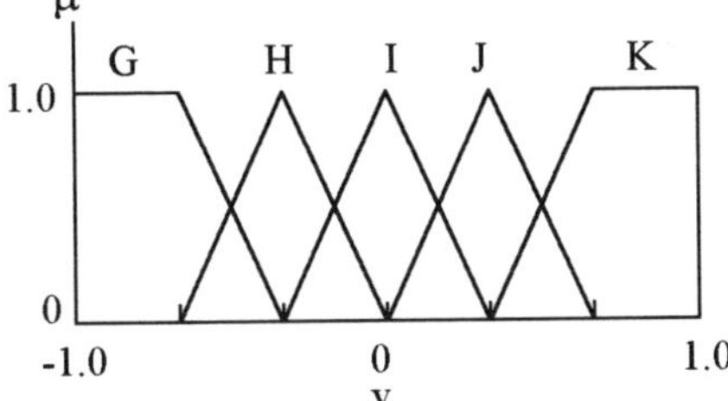

Figure 5.25 Normalised Fuzzy Diagram for the Supervisory Output (v) Sets

A nine rule-base array is shown in Table 5.7, which summarises the fuzzy propositions. The conclusions are aggregated and defuzzified with the use of a fuzzy diagram , such as that shown in Figure 5.26 to yield an associated weight value. It should be noted that the rule-base for the supervisory FLC is to be distinguished from the rule-base for the process FLC (Table 5.8). The objective of the process FLC is to maintain the quality of the product stream within prescribed limits, whilst the objective of the supervisory FLC is to keep the process controller operating around its optimum point. The rules for both these functions would be based upon expert opinion and/or plant operator experience and methods.

Process FLC fuzzy diagrams are illustrated in Figure 5.26 and the corresponding rule-base in Table 5.8. Though an actual controller would usually contain a higher level of partitioning to that shown in Figure 5.26.

Table 5.7 Rule-Base for the Supervisory Output

		e	
u	A	B	C
D	G	H	I
E	H	I	J
F	I	J	K

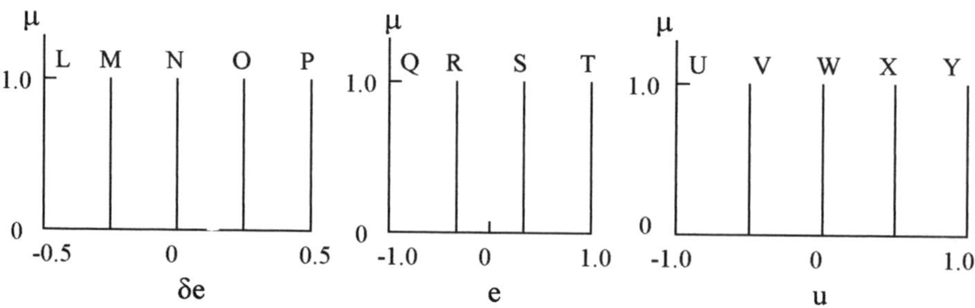

Figure 5.26 Normalised Fuzzy Diagrams for the Process FLC

Table 5.8 Rule Base for the Process FLC Output (u) Sets

e \ **δe**	L	M	N	O
Q	U	U	V	W
R	U	V	W	X
S	V	W	X	Y
T	W	X	Y	Y

As an example of the function of the supervisory controller, consider a process drift that gives a long term mean error of 0.2 and a corresponding controller action of 0.16 (both normalised). Using Figure 5.24, the corresponding membership grades are

$e = 0.2$ $\mu_B = 0.75$ $\mu_C = 0.25$
$u = 0.16$ $\mu_E = 0.8$ $\mu_F = 0.2$

The fuzzy propositions are

IF E	AND U	THEN V	OUTPUT V
B	E	I	0.75 I
B	F	J	0.2 J
C	E	J	0.25 J
C	F	K	0.2 K

At the centroid of the weight sets in the above table the weight factors are

$I = 1$ $J = 1.4$ $K = 1.8$

Aggregating the fuzzy weight factors to give a defuzzified factor

$$v = (0.75*1+1.4(0.2+0.25)+1.8*0.2)/1.4 = 1.243$$

If the initial controller output is u_o and at the current time the input to the process is $u_p = 0.16$, then

$$u_p = vu_o$$

and the current controller output would be u_p/v i.e. 0.129 if the initial state is restored.

An alternative scheme would be one in which the fuzzy rule consequences of the process FLC are each factored by the associated weight (truth) factors, then aggregated as shown in Figure 5.27. The scheme shown in this figure would be appropriate for transient control. The weight factors are deduced from sequential values of the error e and the change of error δe.

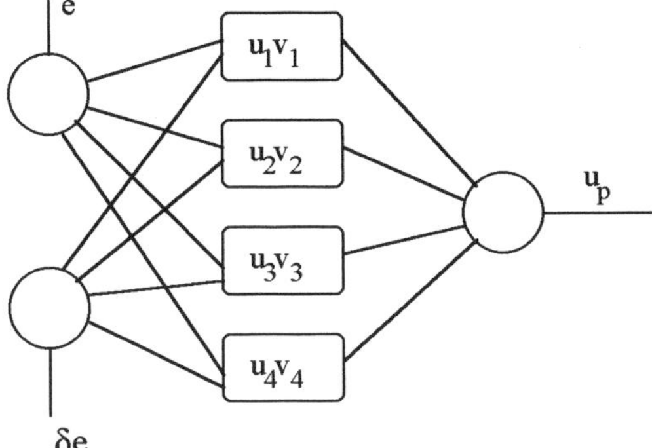

Figure 5.27 Rule Weighting Scheme

Reference may be made to the text by Reznik (1997) for details of FLC parameter adjustment.

5.11 Artificial neural networks (ANN)

ANNs are sometimes linked with FL as a method of determining optimum membership of fuzzy sets. They are also a possible alternative technology for some applications. The creation of this type of network is an attempt to emulate the working of the human brain with its capacity to learn as well as its very large capacity for parallel processing.

The principle of the network is that it comprises an array of interconnected computational elements (CEs), a sample of which is shown in Figure 5.28.

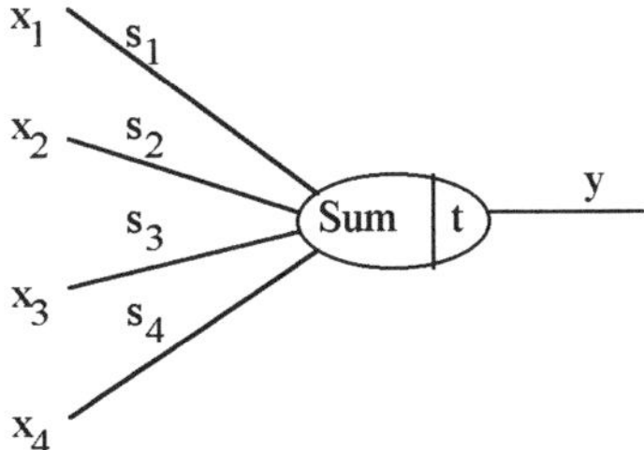

Figure 5.28 A Computational Analogue of a Biological Neuron

The CEs are analogues of biological neurons and are believed to mimic their action. The s_i represents the strength of connection or relation with the preceding nodes.

The overall ANN architecture comprises sheets of CEs that are interconnected. Generally, a CE receives multiple signals via the connections from CEs in preceding layers and has one output to the succeeding layer. This is illustrated in Figure 5.29 for a simple 1*4*1 network. Generally, a CE in the ith layer will be connected to a number of, but not necessarily all, the CEs in the (i+1)th layer. Thus the CE will generally receive a number of input signals as shown in Figure 5.29. These signals are aggregated in some way as indicated by the summation symbol, and compared with a threshold value (t), which if exceeded permits an output signal,y.

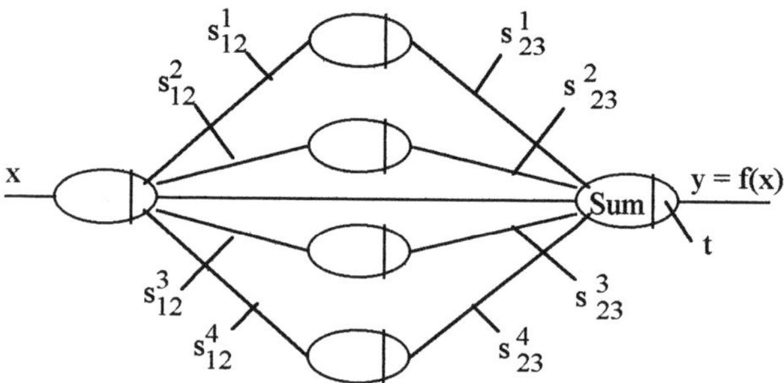

Figure 5.29 A 1*4*1 ANN

There may be many layers of CEs, but only the input and output layers are connected to the environment, the remainder are hidden.

Suppose for example that the signal aggregation is expressed by an arithmetical summation; $s_i x_i = a_n$, then the CE action is given by

$$y = F(a_n - t_n) \qquad \text{for } a_n > t_n \qquad 5.44)$$

and t_n is the given threshold level for a non-zero output. F( ) for this condition is a suitable non-linear function which is usually one of: step function, ramp function or a sigmoidal function as illustrated in Figure 5.30a),b) and c), the choice depending upon the required form of output.

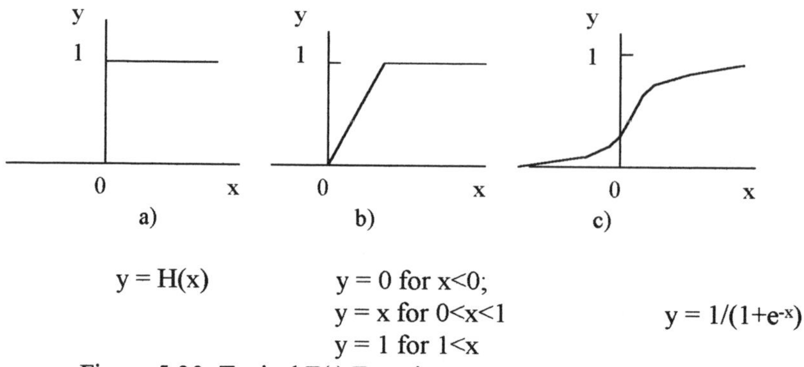

$y = H(x)$

$y = 0$ for $x<0$;
$y = x$ for $0<x<1$
$y = 1$ for $1<x$

$y = 1/(1+e^{-x})$

Figure 5.30 Typical F( ) Functions

Learning is achieved by auto-modification of the strengths of the internal relational strengths by the use of training data. Suppose that a set of training data has sets of given values (x_o, y_o). The input (x_o) data are entered into the ANN (which may be single or multiple input). This will be processed and will result in a set of output data (y). This output data will generally differ from the training output set (y_o), which is the correct set. Hence, there exists an error, ε. Assuming a simple error metric given by

$$\varepsilon = y_o - y \qquad 5.45)$$

where ε is associated with the output CE surface layer, but needs to be distributed back to all layers in the network to achieve learning. (Strengths are always associated with links in an ANN, whereas weights are associated with nodes in a FLC).

The backwards distribution of error is conducted by calculating the ε_{i-1} error from the forward error ε_i by a rule such as

$$\varepsilon_{i-1} = f(x_{k(i-1)}) s_{(i-1)i} \varepsilon_i \qquad 5.46)$$

where $x_{k(i-1)}$ is the kth input to the (i-1)th layer and f( ) is a propagation function such as

$$f(x) = k\exp{-x^2} \qquad \text{(k is a constant)} \qquad 5.47)$$

Then the revised strengths of the link between the (i-1)th and the ith node requires a rule such as

$$s'_{(i-1)i} = s_{(i-1)i} + \delta s_{(i-1)i} \qquad 5.48)$$

where $s_{(i-1)i} = \alpha \varepsilon'_i x_{k(i-1)}$ 5.49)

and $\varepsilon'_i = (\varepsilon_{(i-1)} + \varepsilon_i)/2$ 5.50)

After the first cycle of revisions of the linkage strengths, the trial input data, x_o, are again entered into the ANN and a further cycle is conducted. The correction cycles are repeated until the error falls below an acceptable upper limit. Further sets of training data are entered and compared as described above until all the training data is exhausted and acceptable levels of error are achieved.

Following the training programme, checking data are then entered to confirm reliability and conformance to error limits. The network is then available for application.

CHAPTER 6

INDUSTRIAL ENGINEERING

In general terms, Industrial Engineering is concerned with the creation, organisation and operation of products and the enabling and supporting systems. Thus all systems associated with the production and delivery of goods and services are of interest. The products and systems of production are becoming increasingly well designed and well adapted to their environments, but new demands and priorities and new possibilities of satisfying them are continually being found. Major current applications still lie in the manufacturing and processing industries, as they have in the past, but it is becoming increasingly common to also find applications of industrial engineering methods outside these traditional areas, for example in service areas such as health care. This trend may be expected to continue.

Even in the manufacturing and processing industries the applications encompass a wide range of knowledge such as automation, information technology, operations management, materials handling, quality management, human and financial factors, safety and health and also environmental matters. Within such a wide scope, it is possible in limited space to consider only a selection of topics, other chapters in this text are however also relevant to industrial engineering.

The volume of production of goods or services may be large, in which case the production may be considered to be continuous and is normally product centred, for example, motor car production or petroleum refining both have large annual production volumes, and are product centred. Such production is relatively easy to control as steady state conditions may be approximately achieved, but it is an expensive and relatively inflexible system with high capital investment. Smaller volumes of products are more likely to be treated on a batch production basis, which is characterised by being more difficult to design and control, and is likely to have a less uniform product, but is also less expensive to create and would be more flexible. Continuous processing also lends itself to plant optimisation by the increased use of recycling streams and to the recovery of heat or waste material, therefore the operations have less environmental impact per unit of production (cleaner production).

There is now a widespread global consciousness, which means that systems of all types and generality from the local to the global scale are coming under increased scrutiny, and evaluations which identify and grade all the relevant factors at all levels are increasingly in demand. A knowledge of evaluation methods under conditions of varying degrees of vagueness and uncertainty is therefore required both for producing evaluations and for interpreting the results.

6.1 Process and equipment evaluation

The choice of process or the selection of equipment is an important industrial engineering function, since it will influence the financial and social credibility of an enterprise. A prominent example of this is the exhaustive study by airlines of possible replacement fleet refurbishment aircraft types. In recent years, with increasing globalisation of commercial operations, more diverse and sophisticated equipment and services are becoming available, and this will be further stimulated by the spread of the internet as a communication tool.

Evaluation of equipment, processes and services must account for more than their capital cost. There are also other factors such as performance and controllability, as well as less tangible ones like reliability, maintainability, environmental, health and safety matters to be considered. Evaluation can therefore become a complex and highly subjective issue. Under these circumstances fuzzy logic can provide a credible and useful framework for organising information and quantifying uncertainty in the conclusions, reducing inconsistency, leading to more robust judgements about performance in relation to perceived objectives, thus providing a useful decision making tool.

Evaluation or grading of a system, whether plant, products or services, requires a value judgement. The plausibility of such a value judgement is most readily achieved by comparison with a similar or equivalent concept, for which there is already an acknowledged assessment. Generally, relative merits are easier to ascribe than are absolute values, which is true whether the conceptual parameters are expressed numerically or linguistically.

An important FL method of comparative evaluation is through a composition operation of the object parameters membership values with a given relationship array, which itself defines the reference standard knowledge. A pragmatic example of this is known as bench-marking in quality assurance. The relationship array of the reference standard may be found by a number of different procedures:

i) Cartesian product of the fuzzy membership value vector of the reference system parameters with the fuzzy membership value reference system value judgement.
ii) Classification by similitude or other means.
iii.) Linguistic rule-base in the specific form of a numerical table (known as a look-up table).
iv) Fuzzified closed-form equation.

In case i), let the reference system fuzzy parameters be given by; $X = (x_1,x_2......x_n)$ and the value judgement by; $Y = (y_1,y_2......y_m)$. Then the relationship array (R) of the reference is given in one form by the union of Cartesian products

$$R = (X{\times}Y){\cup}(X'{\times}I) \qquad 6.1)$$

where I is a unit array and X' is the complement of X.

Let the fuzzy parameters of the trial object be the fuzzy set W, then the evaluation of this is given by

$$V = W{\circ}R \qquad 6.2)$$

In membership function notation the procedure is defined by

$$\mu_R = \vee(\mu_x \wedge \mu_y),(1-\mu_x) \qquad 6.3)$$

$$= \begin{vmatrix} \min(x_1,y_1) & \min(x_1,y_2) \ldots\ldots\ldots\ldots & \min(x_1,y_m) \\ . & . \quad \ldots\ldots\ldots\ldots & . \\ . & . \quad \ldots\ldots\ldots\ldots & . \\ . & . \quad \ldots\ldots\ldots\ldots & . \\ \min(x_n,y_1) & \min(x_n,y_2) \ldots\ldots\ldots\ldots & \min(x_n,y_m) \end{vmatrix} \cup \begin{vmatrix} 1-x_1 & 1-x_1 \ldots\ldots\ldots & 1-x_1 \\ . & . \ldots\ldots\ldots & . \\ . & . \ldots\ldots\ldots & . \\ . & . \ldots\ldots\ldots & . \\ 1-x_n & 1-x_n \ldots\ldots\ldots & 1-x_n \end{vmatrix}$$

or

$$\mu_R = \begin{matrix} \\ x_1 \\ . \\ \\ . \\ x_n \end{matrix} \begin{vmatrix} y_1 & y_2 \ldots\ldots\ldots & \ldots y_m \\ r_{11} & r_{12} \ldots\ldots\ldots\ldots & r_{1m} \\ . & . \ldots\ldots\ldots\ldots\ldots & \\ . & \ldots\ldots\ldots\ldots\ldots & \\ . & . \ldots\ldots\ldots\ldots\ldots & \\ r_{n1} & r_{n2} & \ldots r_{nm} \end{vmatrix} \qquad 6.4)$$

Also, $\mu_V = \vee \mu_W \wedge \mu_R \qquad 6.5)$

where

$$\begin{aligned} \mu_{v1} &= \max(\min(w_1,r_{11}),\min(w_2,r_{12}) \ldots\ldots \min(w_m,r_{1m})) \\ \mu_{v2} &= \max(\min(w_1,r_{21}),\min(w_2,r_{22}) \ldots\ldots \min(w_m,r_{2m})) \\ . & \quad \ldots\ldots\ldots\ldots\ldots\ldots\ldots\ldots\ldots\ldots\ldots\ldots \\ . & \quad \ldots\ldots\ldots\ldots\ldots\ldots\ldots\ldots\ldots\ldots\ldots\ldots \\ \mu_{vn} &= \max|(\min(w_1,r_{n1}),\min(w_2,r_{n2}) \ldots\ldots \min(w_m,r_{nm})) \end{aligned} \qquad 6.6)$$

The μ_{vi} give the membership function values of the value judgement.

Example 6.1

An industrial organisation is considering updating a process plant with a view to improving its commercial position. The present plant has given satisfactory results for a number of years, but requires refurbishment. The alternative is to purchase new plant which incorporates more recent technology. Both a technical and a financial appraisal are required, but only the technical appraisal is to be considered here.

The management has identified four major performance parameters:
i) Plant productivity
ii) Product quality
iii) Plant reliability
iv) Technical support

The merits of the plant proposal are to be considered in four categories:
i) Superior (S)
ii) Good (G)
iii) Moderate (M)
iv) Inferior (I)

It is estimated that after refurbishment the present plant will have the following profile:

A = [0.5//P+0.6//Q+0.9//R+0.4//T]

The management team also reports a consensus value judgement of this profile as

B = [0.2//S+0.4//G+0.6//M+0.2//I]

Expert opinion is that the proposed new plant performance is likely to have the following profile

C = [0.6//P+0.7//Q+0.7//R+0.9//T]

Estimate the value judgement of the proposed new plant.

Solution

First it is necessary to construct the fuzzy relational membership array, from the data supplied for the refurbished existing plant. This will provide the reference standard, which is evaluated as follows:

$$\min(\mu_A,\mu_B) = \begin{vmatrix} 0.5 \\ 0.6 \\ 0.9 \\ 0.4 \end{vmatrix} \begin{vmatrix} 0.2 & 0.4 & 0.6 & 0.2 \end{vmatrix} = \begin{vmatrix} 0.2 & 0.4 & 0.5 & 0.2 \\ 0.2 & 0.4 & 0.6 & 0.2 \\ 0.2 & 0.4 & 0.6 & 0.2 \\ 0.2 & 0.4 & 0.4 & 0.2 \end{vmatrix}$$

$$\min((1-\mu_\alpha),1) = \begin{vmatrix} 0.5 \\ 0.4 \\ 0.1 \\ 0.6 \end{vmatrix} \begin{vmatrix} 1 & 1 & 1 & 1 \end{vmatrix} = \begin{vmatrix} 0.5 & 0.5 & 0.5 & 0.5 \\ 0.4 & 0.4 & 0.4 & 0.4 \\ 0.1 & 0.1 & 0.1 & 0.1 \\ 0.6 & 0.6 & 0.6 & 0.6 \end{vmatrix}$$

Also,

$$\mu_R = \max(\min(\mu_A,\mu_B)\min(1-\mu_A),1))$$

$$= \begin{vmatrix} 0.2 & 0.4 & 0.5 & 0.2 \\ 0.2 & 0.4 & 0.6 & 0.2 \\ 0.2 & 0.4 & 0.6 & 0.2 \\ 0.2 & 0.4 & 0.4 & 0.2 \end{vmatrix} \vee \begin{vmatrix} 0.5 & 0.5 & 0.5 & 0.5 \\ 0.4 & 0.4 & 0.4 & 0.4 \\ 0.1 & 0.1 & 0.1 & 0.1 \\ 0.6 & 0.6 & 0.6 & 0.6 \end{vmatrix}$$

The relationship array is therefore

$$\mu_R = \begin{vmatrix} 0.5 & 0.5 & 0.5 & 0.5 \\ 0.4 & 0.4 & 0.6 & 0.4 \\ 0.2 & 0.4 & 0.6 & 0.2 \\ 0.6 & 0.6 & 0.6 & 0.6 \end{vmatrix}$$

Let V be the value judgement of the proposed new plant, then

$$\mu_V = \max(\min(\mu_C,\mu_R))$$

or

$$\mu_V = \begin{vmatrix} 0.6 & 0.7 & 0.7 & 0.9 \end{vmatrix} \begin{vmatrix} 0.5 & 0.5 & 0.5 & 0.5 \\ 0.4 & 0.4 & 0.6 & 0.4 \\ 0.2 & 0.4 & 0.6 & 0.2 \\ 0.6 & 0.6 & 0.6 & 0.6 \end{vmatrix}$$

$$= \begin{vmatrix} 0.6 & 0.6 & 0.6 & 0.6 \end{vmatrix}$$

Therefore the evaluation of the new plant value judgement is

$$V = [0.6//S+0.6//G+0.6//M+0.6//I]$$

This value judgement is an improvement in the superior and good categories, but is counterbalanced by a higher level in the inferior category. The outcome would need to be linked with the corresponding financial analysis of the project to reach an overall conclusion.

6.2 Resource assignment

In the manufacturing and processing industries as well as in other major sectors such as commercial services, health services, utilities and the armed forces, the optimum allocation of resources is an important consideration. In operational research the subject is viewed as a sub-class of the general topic of transportation and is treated in several different ways (for example, by the so-called Hungarian method, which is described in standard texts). Because of the divers interest in resource allocation the problem arises in different guises.

Resource allocation is important in the planning of physical facilities associated with the product or service offered, including all aspects of the operations such as storage of materials and the functioning of support services. General and detailed layout of facilities are affected, including not only processing facilities, but also materials-handling equipment, utilities and building construction.

The two distinctive forms of physical resources layouts are the product focused and the process focused. The former is suited to mass standardised production and the latter to high variety lower volume demands. The optimum matching of resources and demands, or donors and beneficiaries in the case of aid projects, is a basic problem in industrial engineering. This is discussed in more detail in the following section.

6.3 Multiple resources and demands (Donors and beneficiaries)

The problem of matching resources and demands can arise in a number of different ways. Consider the case of one-to-one matching of resources and demands, where the resource servers are non-specialised. This can of course be accomplished in a number of different ways, but optimum matching requires a criterion to be defined which can be the optimising subject. A typical class of this type is shown in Figure 6.1.

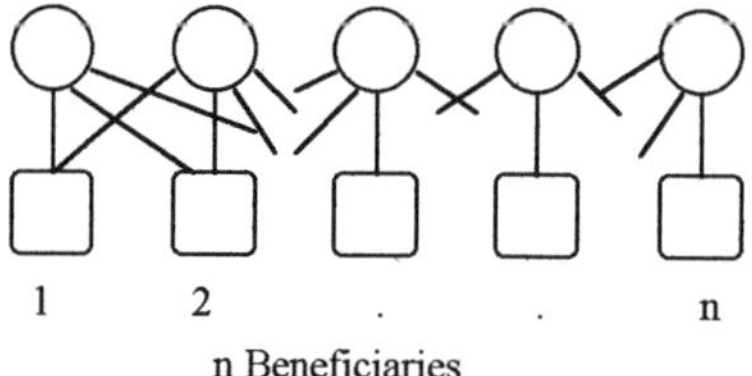

Figure 6.1 The Case of 1:1 Resources and Beneficiaries

The beneficiaries create specific demands and the sources serve the resources specifically required. If the demands of the n beneficiaries and the services of the n resources are each identical, then there would obviously be n^2 ways of making the relationships, each being identical, no optimisation is then possible. However, if the services provided by the resources were all different (as is the case for different machine-operator pairs or aid organisations for example) then the n^2 relations would all be different, and there would now be a case for optimisation, as there would be for the cases between the two extremes.

Whilst the treatment of this problem may be found in standard operational research texts, it is also possible to find one using similitude, as shown in the example below. This solution is simpler and more direct than the alternatives.

Example 6.2

In a manufacturing industry, four different types of component are required for different process operations which can be supplied by each of four machine-operator units. From work studies it is known that the times taken for each of the four operations are different for each of the machine-operator units. The data is summarised in Table Ex. 6.1 below, where the entries represent the time taken in minutes.

Table Ex. 6.1 Component and Machine-Operator Times

		Component			
		1	2	3	4
Machine-	1	17	11	13	15
Operator	2	12	16	14	9
	3	10	15	13	6
	4	19	16	15	9

Find the optimum allocation of machine-operator to components to minimise the total time required.

Solution

By a similitude analysis based upon the min/max method, arrays of similitude metrics may be found for machine-operators (rows) and components (columns). For example, the similitude metric for row 2 and row 3 is

$$\mu_{23} = (\min(12,10)+\min(16,15)+\min(14,13)+\min(9,6))/\\(\max(12,10)+\max(16,15)+\max(14,13)+\max(9,6))$$

$$\mu_{23} = (10+15+13+6)/(12+16+14+9)\\ = 0.86$$

and clearly $\mu_{23} = \mu_{32}$

Thus the similarity by rows (machine-operator) metric array is given by

$$R = \begin{matrix} 1 & 0.72 & 0.79 & 0.67 \\ & 1 & 0.86 & 0.56 \\ & \text{sym} & 1 & 0.65 \\ & & & 1 \end{matrix}$$

Also by the same procedure, the similarity by columns (components) metric array is given by

	1	0.73	0.79	0.67
C =		1	0.79	0.67
		sym	1	0.64
				1

The smallest values of the similarity metrics in the above two arrays give the pairs of machine-operator or components, as the case may be, which are most dissimilar. Likewise, the largest values give the pairs which are most similar.

Considering the R-array and listing the most dissimilar pairs

R-array	**Pairs**	**Similarity metric**
	(4,2)	0.56
	(4,3)	0.65
	(4,1)	0.67

The most similar pairs are therefore

(3,2)	0.86
(3,1)	0.79

Considering next the S-array and listing the most dissimilar pairs

S-array	**Pairs**	**Similarity metric**
	(4,3)	0.64
	(4,2)	0.67
	(4,1)	0.67

The most similar pairs are therefore

(3,1)	0.79
(3,2)	0.79

The similarity metrics may now be used to order the machine-operator and the component elements. The R-array metrics show that the most dissimilar machine-operator elements are 4 and 2. They may therefore be placed at the extremes of the list of elements, thus: 4.....2. The most similar pairs are 3 and 2, thus giving, 4....3,2. There remains only the one space to fill; thus the ordering of these elements is: 4,1,3,2.

Treating the S-array metrics in a similar way it is noted that the most dissimilar components elements are 4 and 3, thus the ordering is: 4.......3. The strongest similarities are (3,1) and (3,2), which are both equal. Element 1 and element 2 have equal dissimilarity strengths to element 4, therefore either of the following two orders are possible: 4,1,2,3 or 4,2,1,3.

Consider now matching the machine-operator and component elements, the following combinations are possible:

Machine-operator	**Component** or	**Component**		**Time** (mins)	
4	4	3		9 or	15
1	1	2		17	11
3	2	1		15	10
2	3	4		14	9
			$\Sigma =$	55	45

Machine-operator	**Component** or	**Component**	**Time** (mins)	
4	4	3	9 or	15
1	2	1	11	17
3	1	2	10	6
2	3	4	14	9
		$\Sigma =$	44	56

It may be noted that there are two minimum summations: 44 and 45 mins. They differ by a negligible amount, certainly within the error of working measurement, and are therefore sensibly the same. Hence the ambiguity in the ordering of the S-array metrics.

The other two values of 55 and 56 are the maximum times, and they too differ by a negligible amount.

The solution to the problem may be expressed as a diagonal array

		Component			
		4	1	3	2
Machine-	4	9			
operator	2		11		
	1			10	
	3				14

This array expresses the optimum machine-operator and component pairs in a convenient form.

In the above example, whether the shorter contact time between machine-operator 4 and component 4 compared with other combinations could be used to increase the number of other components would of course depend upon the change-over time required, and also on the relative volume demand for the other components.
Overall however, the total time has been minimised and this means that the work in progress for a given level of production is reduced compared with any of the other possible combinations, (except for the ambiguous case). The processing time from input to output is reduced, which makes possible a smaller inventory of finished products.

6.4 Clustered resources and demands

As mentioned previously, physical resources planning is governed by a preference for a product or a process bias, depending upon the nature of the product, whether it is uniform and large volume or high variety small volume. There are of course gradations between these extremes, and for intermediate patterns of demand there is a system called Group Technology, which confers some of the benefits of both the product and process centred types of operations. In this system, like elements of resources and demands are clustered together into semi-production cells. This is applicable to moderate volumes of production of a restricted variety of outputs. The advantage over a totally process based system is that materials handling is simplified, again leading to reduced work in progress and reduced inventory.

Various methods of resource/ beneficiary clustering are known, the treatment outlined below is based on a similarity principle which is a flexible and convenient framework for many related problems.

Let $(x_1,x_2........x_m)$ be the set of resource elements and $(y_1,y_2........y_n)$ be the set of beneficiary elements. Then the relations between the x_i and y_j may be represented by an nxm array as follows:

Table 6.1 Typical Resource-Demand Relational Array

	x_1	$x_2........x_m$
y_1	0	1.........0
y_2	0	0.........1
.	.	. .
.	.	. .
y_n	1	0..........1

Row and column similarity treatments, such as those carried out previously, yield an n^2 array and an m^2 array respectively

$$R = \begin{matrix} r_{11} & r_{12}........r_{1n} \\ . & . \quad\quad . \\ . & . \quad\quad . \\ r_{n1} & r_{n2}........r_{nn} \end{matrix} \qquad S = \begin{matrix} s_{11} & s_{12}........s_{1m} \\ . & . \quad\quad . \\ . & . \quad\quad . \\ s_{m1} & s_{m2}.......s_{mm} \end{matrix} \qquad 6.7)$$

The above similitude arrays are symmetrical; rij = r_{ji} and $s_{ij} = s_{ji}$. Also $r_{ii} = s_{ii} = 1$ (the arrays are reflexive). The r_{ij} and s_{ij} $(i \neq j)$ may in general, take on values in the range, $0 \leq r \leq 1$ and $0 \leq s \leq 1$. Also, in general the $r_{ij}(s_{ij})$ will exhibit only a tolerance relation (reflexivity and symmetry) and not equivalence. That is

$$r_{ik} = \lambda_a; \; r_{kj} = \lambda_b \;\text{ and } r_{ij} = \lambda < \min(\lambda_a, \lambda_b) \qquad 6.8)$$

Equations 6.8) state that the $r_{ij}(s_{ij})$ generally do not have the property of transitivity.

By inspection of the R and S arrays in equations 6.7) the r_{ij} and s_{ij} may be listed in order of increasing (or decreasing) orders of similarity. The R and S clusters may then be related to each other by reference to the raw data in Table 6.1. The following example will clarify the procedure.

Example 6.3

In an animal feed processing plant eight feedstocks require to be processed in unit operations. The feedstock/operations pattern is shown in the following table of data.

Table Ex. 6.2 Feedstock/Unit operations Pattern

		Feedstock							
		1	2	3	4	5	6	7	8
Unit operation	1	1	1			1			
	2				1				1
	3		1	1			1	1	
	4				1				1
	5				1		1	1	
	6	1	1			1			

Determine the most efficient feedstock/unit operations cell clustering pattern.

Solution

The data given in Table Ex. 6.2 is analysed for similarities by the max/min procedure. The similarity elements are defined by

$$r_{ij} = \Sigma_{k=1}{}^{n}\min(y_{ik},y_{jk})/\Sigma_{k=1}{}^{n}\max(y_{ik},y_{jk})$$

Similarly, $s_{ij} = \Sigma_{k=1}{}^{m}\min(x_{ik},x_{jk})/\Sigma_{k=1}{}^{m}\max(x_{ik},x_{jk})$

The results are shown below in Table Ex. 6.3 and Table Ex 6.4

Table Ex. 6.3 Similarity by Rows for Operations

	1					
	0	1				
	1/6	0	1	Sym.		
R =	0	1	0	1		
	0	1/5	3/5	1/5	1	
	1	0	1/6	0	0	1

Table Ex. 6.4. Similarity by Columns for Feedstock

1							
2/3	1						
0	0	1					
0	0	1/4	1	Sym.			
1	1	0	0	1			
0	1/4	1	1/4	0	1		
0	1/4	1	1/4	0	1	1	
0	0	0	2/3	0	0	0	1

From Table Ex. 6.3 the following list of similarity rankings may be obtained in decreasing order:

Unit operations pair similarity:
(1,6)(2,4) = 1
(3,5) = 3/5
(2,5)(4,5) = 1/5 (Not required)
(1,3)(3,6) = 1/5 (Not required)

The last two entries in the above list are not required because the higher order rankings exhaust the unit operations list.

From Table Ex. 6.4 the following list of similarity rankings may be obtained in decreasing order

Feedstock pair similarity: (1,5)(2,5)(3,6)(3,7)(6,7) = 1
(4,8) = 2/3

(Feedstock list exhausted).

It may be noted that feedstock 5 is common to the first pair in line 1 of the above list, also 3,6 and 7 are common to the third , fourth and fifth pairs, hence the clusters may be condensed to the following without loss of similarity

$$(1,2,5)(3,6,7) = 1$$
$$(4,8) = 2/3$$

The feedstock and unit operations groups may now be related together by consideration of the data in Table Ex. 6.2

Feedstock Group Unit Operations Group

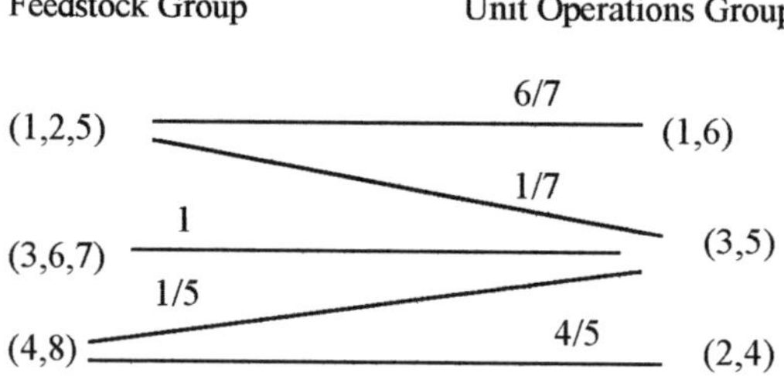

Figure Ex. 6.2 Feedstock-Unit Operations Group Relations

Figure Ex. 6.2 shows the strengths of the links between the feedstock and the unit operations groups. For example, feedstock group (1,2,5) makes a total of seven demands on the unit operations (1,6,3) and of these, six are on 1 and 6, and one is on operation 3, making strengths of 6/7 and 1/7 respectively with groups (1,6) and (3,5).

The clustering may also be portrayed in the form of a diagonalised chart as shown in the following tabulation:

Table Ex. 6.5 Diagonalised Pattern of the Feedstock/Unit Operations Relations

	Feedstock							
	2	1	5	3	6	7	4	8
Opns.								
1	1	1	1					
6	1	1	1					
3	1			1	1	1		
5				1	1	1	1	
2							1	1
4							1	1

It may be noted in the above example, that without feedstock 2 demand for operation 3, and without a feedstock 4 demand for operation 5, perfect diagonalisation would be achieved. This would mean that the feedstock and unit operations groups would be mutually exclusive. In this case, the similarity tables (Tables Ex. 6.2 and 6.3) would contain only binary entries (0,1). This represents the case of complete similarity or none. Fractional numbers represent a degree of fuzziness and imperfect cells are formed.

6.5 Clustering very fuzzy systems

Fractional off diagonal entries in the similarity arrays represent a degree of fuzziness and this results in imperfect cells being formed and imperfect diagonalisation of the data, as exemplified in the previous example, which is only mildly fuzzy. When the data is very fuzzy, then a simple similarity analysis may be unable to provide sufficient information about any underlying fuzzy structure. Such structures may be exposed however by composition of the data.

Generally, the similitude arrays for rows and columns of the raw data array will exhibit only tolerance relations and in this case it may be found that groups formed by arranging the data in decreasing order of similarity will not exhaust all the elements, if the system is very fuzzy. This problem may be overcome by a composition operation of the data defined by

$$R^2 = RoR; \qquad R^3 = RoR^2 \qquad \text{etc} \qquad 6.9)$$

Similarly for the S array

Max/min composition is defined by

$$r_{ij}^2 = \vee r_{ik} \wedge r_{kj} \qquad \text{(Max over repeated subscripts).} \qquad 6.10)$$

Consider for example a very fuzzy 7x5 array created by a random selection process, as shown below in Table 6.2

Table 6.2 Random 7x5 Binary Array

Resources \ **Beneficiaries**	1	2	3	4	5	6	7
1	1	1	1		1		
2			1	1	1		
3	1			1		1	
4				1		1	
5		1	1				1

A similarity study of this data by rows and columns yields the following arrays

Table 6.3 Similitude Array by Rows

R =	1	2	3	4	5
1	1				
2	2/5	1	Sym.		
3	1/6	1/5	1		
4	0	1/4	2/3	1	
5	2/5	1/5	0	0	1

Table 6.4 Similitude Array by Columns

S =	1	2	3	4	5	6	7
1	1						
2	1/3	1					
3	1/4	2/3	1	Sym.			
4	1/4	0	1/5	1			
5	1/3	1/3	2/3	1/4	1		
6	1/3	0	0	2/3	0	1	
7	0	1/2	1/3	0	0	0	1

Let λ be a given similitude cut-off level in the R and S arrays. For the R-array, by inspection of Table 6.3, the groups associated with selected values of λ are listed below.

λ Value	**R Group**
2/3	(3,4)
2/5	(1,2)(1,5) or (1,2,5)

λ Value	**S Group**
2/3	(2,3)(3,5)(4,6) or (2,3,5)(4,6)
1/2	(2,7)
1/3	(1,2)(1,5)(1,6)(3,7)(2,5) or (1,2,5,6)(3,7)

It may be noted that since the elements 2,3,5 and 6 are all accounted for at the $\lambda_{2/3}$ level, they may be deleted at the $\lambda_{1/2}$ and $\lambda_{1/3}$ levels. In this case, the S-array simplifies as follows:

λ Value	**S Group**
2/3	(2,3,5)(4,6)
1/3	(7)
1/3	(1)

The R and S clusters may be obtained in another way. By composition of the arrays in Tables 6.3 and 6.4 using equations 6.9) and 6.10), the following arrays for R and S respectively are found:

Table 6.5 The R^2 Array

		1	2	3	4	5
	1	1				
	2	2/5	1	Sym.		
$R^2 =$	3	1/5	1/4	1		
	4	1/4	1/4	2/3	1	
	5	2/5	2/5	1/5	1/5	1

Table 6.6 The S^2 Array

	1	2	3	4	5	6	7
1	1						
2	1/3	1					
3	1/3	2/3	1	Sym.			
4	1/3	1/4	1/4	1			
5	1/3	2/3	2/3	1/4	1		
6	1/3	1/3	1/4	2/3	1/3	1	
7	1/3	1/2	1/2	1/5	1/3	0	1

Considering the R-array in Table 6.5, the maximum cut-off level now needs to be chosen such that every R element is included. This is found by inspection to be at $\lambda = 2/5$. Similarly, for the S-array in Table 6.6 the maximum cut-off level to include all elements is $\lambda = 2/3$. The corresponding membership patterns are shown in Tables 6.7 and 6.8 below.

Table 6.7 The R Membership Pattern.

	1	2	3	4	5
1	1	1			1
2	1	1			1
3			1	1	
4			1	1	
5	1	1			1

Table 6.8 The S Membership Pattern.

	1	2	3	4	5	6	7
1	1						
2		1	1		1		
3		1	1		1		
4				1		1	
5		1	1		1		
6				1		1	
7							1

From Tables 6.7 and 6.8 the following groups are obvious

R	**S**
(1,2,5)	(2,3,5)
(3,4)	(4,6)
	(1)
	(7)

It will be found that these membership patterns do not exist at the same cut-off values in the R and S arrays, Tables 6.3 and 6.4 respectively. Therefore the effect of composition is to sharpen the similitude membership patterns.

The R and S clusters may be related together as shown in Figure 6.2.

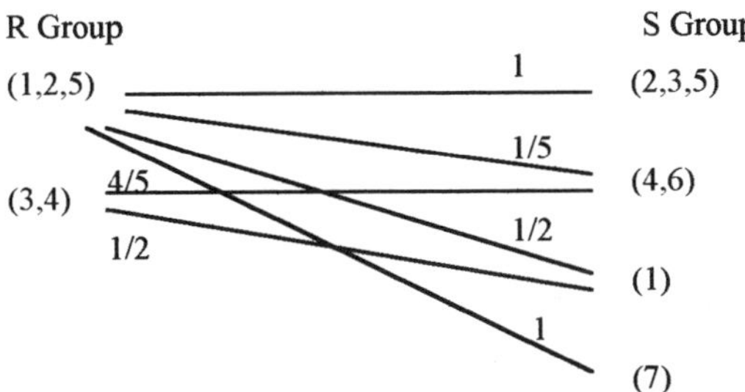

Figure 6.2 Fuzzy R and S Group Relationships

Obviously, the S group (7) could be combined with the S group (2,3,5), thereby reducing the number of S groups to three. The corresponding fuzzy diagonalisation is shown in Table 6.9.

Table 6.9 Diagonalisation Pattern of the Random Data Array

R \ S	2	3	5	7	1	4	6
5	1	1		1			
1	1	1	1		1		
2		1	1			1	
3					1	1	1
4						1	1

It is worth noting that a certain degree of structure exists even with initially random data of a limited extent. The implication is that there is merit in clustering a limited number of beneficiaries and resources even though they may appear to be divers in type.

It will be recalled that in the similitude arrays, binary entries represent complete similarity (1) or none(0) between rows (or columns as the case may be). Fractional entries represent fuzzy membership values, which in turn imply fuzzy relationships between groups. A correlation metric may be used to assess the degree of fuzziness, β, (see the Appendix) of a set or an array. This may be defined as

$$\beta = a/b \tag*{6.11)}$$

where a and b are the intersection and union respectively of the array and its complement. For the entries in Tables 6.3 and 6.4 β = 0.1848 and 0.3447 respectively. Dividing each value by the number of entries in the corresponding Table gives 0.007382 and 0.007181 respectively, which are similar values.

Clustering or grouping has a wide variety of applications in industrial engineering and is part of the wider study of pattern recognition. In industrial engineering the applications are not only process studies but also in product inventories, assembly operations, inspection, testing, tooling and of course in organising data banks.

6.6 Work study

The basic objective of work study is to make the most effective use of the resources found in all forms of human activity, whether classified as work or leisure, such as sport. As may be anticipated, the concept of human effort features prominently in this. The general subject of work study is subdivided into two quite distinct, but closely related areas; method study, which is concerned with procedures, and work measurement, which is concerned with work content. This is illustrated in Figure 6.3. (The British Standard BS 3138:1992 may be consulted for definitions of terms).

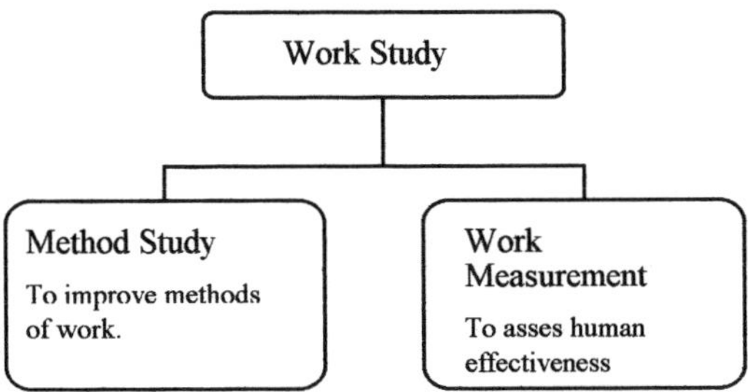

Figure 6.3 The Sub-Divisions of Work Study

The benefits accruing from methods study and work measurement include: more effective use of all resources, including manpower, equipment, natural resources and the environment, improved planning and also the control of human activities, though this must be allied with an understanding of human nature.

In the more industrialised countries work study has attracted considerable interest during the last half century, largely because of the intensifying local national and global economic competition for satisfaction, conceived as the beneficiaries. It has therefore become an increasingly important management tool, particularly with the growth in the use of the internet and the networking of activities. Most people practise work study in a conscious or unconscious way, for example by taking short cuts wherever the opportunity presents itself and will show an economy of time, effort or money. Fewer people however persue this matter in a systematic and rational way.

6.6.1 Method study

Work study methods may be applied to all scales of activity from eye movements of aircraft pilots to the operation of international cargo transport. One of the possible benefits is found in work standardisation, and as such it is an essential precursor to work measurement. Standardisation of equipment and working conditions is also required. However too much personal standardisation of effort can lead to monotony and to psychological problems.

Group technology finds an application in work standardisation in several ways. It is useful in identifying the degree of commonality between features of divers products and activities. This information may be useful for production planning purposes, for example, for assembling batches of different components with similar features for similar manufacturing processes. Dissimilar features may then be processes individually downstream, and possibly new batches may be formed which are base on other similar features. This is illustrated in Figure 6.4.

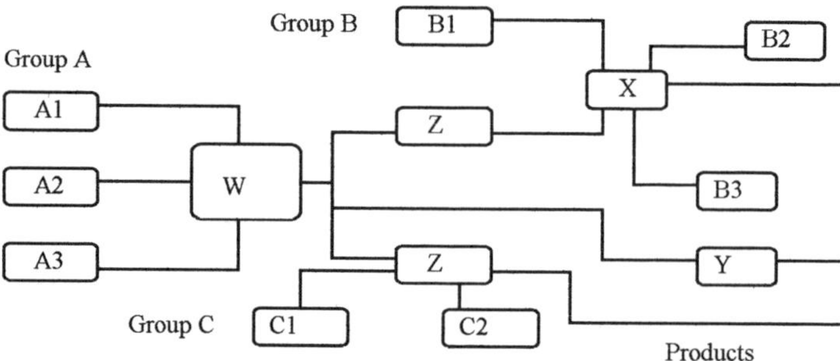

Figure 6.4 Successive Clustering of Process Products
X,Y and Z are Processes

In Figure 6.4, Groups A,B and C are the inputs to the processes X,Y and Z. Downstream clustering of process products occurs as show.

Computer numerically controlled (CNC) processes can be an alternative to some of the advantages of clustering in that they are able to swiftly adapt to (limited) new demands, but CNC technology is not universally applicable. Clustering occurs quite frequently in the chemical process industries where, for example, successive blending of various feedstocks permits a range of product streams.

Example 6.4

A clothing manufacturer is planning a production run of a new range of garments for a future season. The garment range is designed for the better quality, moderate volume part of the market. It is planned to manufacture the basic garment in several different fashionable colours and to impart variety in the finished goods by selecting batches from the basic garment production for different standard finishes, which will include various trims and accessories.

The garment designers plan to divide the basic garments into five batches (b_i)and to apply selected finishes (f_i) as shown in Table Ex. 6.6.

Table Ex. 6.6 Schedule of Garment Finishes

	Batch				
Finish	b_1	b_2	b_3	b_4	b_5
f_1	1	0	0	1	1
f_2	1	0	1	0	1
f_3	0	1	0	1	1
f_4	1	0	1	1	0
f_5	0	0	1	0	0

Produce a production flow schedule.

Solution

The production pattern can be arranged such that the garments are grouped to receive the standard finishes in successive stages according to the schedule provided in Table 6.6. The grouping can be achieved by applying similarity analysis to Table 6.6 to achieve the similarity by rows (F-array) and similarity by columns (B-array) shown in Table Ex. 6.7 and Table Ex. 6.8.

Table Ex. 6.7 Similarity by Rows

1	0.25	0.5	0.5	0
	1	0.2	0.5	0.33
		1	0.2	0
	Sym.		1	0.33
				1

Table 6.8 Similarity by Columns

1	0	0.5	0.5	0.5
	1	0	0.25	0.33
		1	0.2	0.2
		Sym.	1	0.5
				1

Choosing descending similarity cut-off levels (λ), the following groups are found:

λ Cut-Off Level	**F-Group**	**B-Group**
1/2	(1,3)(1,4)(2,4) = (1,3)(2,4)	(1,3)(1,4)(1,5) = (1,3)(4,5)
1/3	(2,5)(4,5) = (5)	(2,5) = (2)
	(All f_i exhausted)	(All b_i exhausted)

Note that where the f_i or b_i appear at higher values, they may be deleted.

On the basis of the above similarity analysis the schedule of garment batch finishes may be arranged as shown below in Table Ex. 6.9.

Table 6.9 Clustered Garment Finish Schedule

	b_1	b_2	b_3	b_4	b_5
f_1	1	1	1		
f_2	1	1			1
f_3	1		1	1	
f_4		1	1	1	
f_5				1	

Evaluating the F and B arrays using equation 6.9)

$$F^2 = F \circ F \qquad \text{and } R^2 = R \circ R$$

The arrays in Table Ex. 6.10 and Table Ex. 6.11 are found.

Table Ex. 6.10 The F^2 Array

$F^2 =$

1	0.36	0.51	0.44	0.17
	1	0.26	0.59	0.40
		1	0.33	0.13
	Sym.		1	0.31
				1

Table Ex. 6.11 The B^2 Array

$B^2 =$

1	0.17	0.47	0.62	0.51
	1	0.17	0.26	0.28
		1	0.34	0.33
		Sym.	1	0.64
				1

At the cut-off level of 0.51 the membership arrays are as shown in Table Ex. 6.12 and Table Ex. 6.13.

Table Ex. 6.12 Membership for the F Array

$F_{0.51}{}^2 =$

1	2	3	4	5
1	.	1	.	.
.	1	.	1	.
.	1	.	1	.
.	1	.	1	.
.	.	.	.	1

Table Ex. 6.13 Membership for the B Array

$B_{0.51}{}^2 =$

1	2	3	4	5
1	.	.	1	1
.	1	.	.	.
.	.	1	.	.
1	.	.	1	1
1	.	.	1	1

Comparing the results in Tables Ex. 6.12 and 6.13 with the previous F Groups and B Groups, the clusters are clearly distinguishable in these tables and the following groups are identified

F Group.	**B Group**
(1,3)(2,4)(5)	(1,4,5)(2)(3)

The above analysis indicates that the production flow schedule should be as shown in Figure Ex. 6.3.

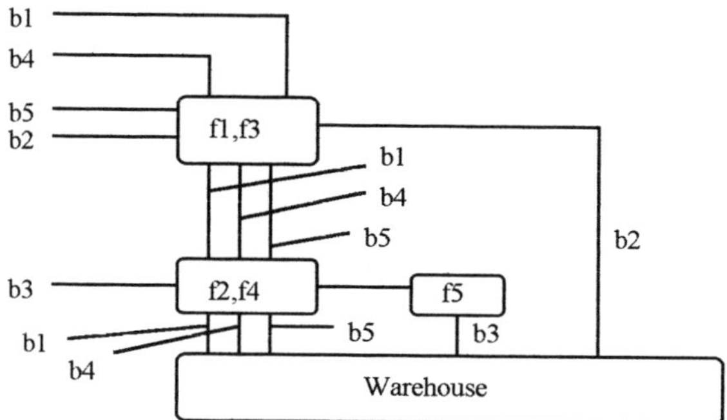

Figure Ex. 6.1 Proposed production Flow Chart for Garment Finishing

Work study should commence at the operations planning phase and should continue into the actual operations phase. This is initially done by breaking unit operations into elements for detailed study using product and process drawings, models prior data where relevant, computer simulation and expert knowledge. Then later, by study of the actual operations. Motion studies however conducted may lead to effective combinations, to shortening paths or complete elimination of some operations. The benefit being that of conservation of effort for a given effect.

Example 6.5

A large national bank is evaluating staff service conditions in its branches to compare with the variable customer service complaints received during a twelve month survey period. As a result of a staff questionnaire, the management board have chosen a number of service conditions considered to be most important in affecting the staff work performance, these are: salary level (L), technical support (T), management support (S) and number of personal benefits (N). The bench-mark service conditions are represented by a fuzzy set, A, where

$$A = [0.9//L+0.8//T+0.7//S+ 0.7//N]$$

A value judgement on the service conditions would be given with grades, for example; superior (SU), standard (ST), low standard (LS) and inferior (IN).

The value judgement (B) corresponding with these service conditions is

$$B = [0.6//SU+0.8//ST+0.6//LS+0.3//IN]$$

Deduce the value judgement of a branch of the bank which has the following service conditions: L = 0.7, T = 0.6, N = 0.4 and S = 0.3.

Solution

A relationship array between the given service conditions and the corresponding value judgement using equation 6.3).Now in this expression

$$\mu_x \wedge \mu_y = \begin{vmatrix} 0.9 \\ 0.8 \\ 0.7 \\ 0.7 \end{vmatrix} \times \begin{vmatrix} 0.6 & 0.8 & 0.6 & 0.3 \end{vmatrix} = \begin{vmatrix} 0.6 & 0.8 & 0.6 & 0.3 \\ 0.6 & 0.8 & 0.6 & 0.3 \\ 0.6 & 0.7 & 0.6 & 0.3 \\ 0.6 & 0.7 & 0.6 & 0.3 \end{vmatrix}$$

and

$$(1-\mu_x) = \begin{vmatrix} 0.1 \\ 0.2 \\ 0.3 \\ 0.3 \end{vmatrix} \times \begin{vmatrix} 1 & 1 & 1 & 1 \end{vmatrix} = \begin{vmatrix} 0.1 & 0.1 & 0.1 & 0.1 \\ 0.2 & 0.2 & 0.2 & 0.2 \\ 0.3 & 0.3 & 0.3 & 0.3 \\ 0.3 & 0.3 & 0.3 & 0.3 \end{vmatrix}$$

Now $\mu_R = \vee((\mu_x \wedge \mu_y),(1-\mu_x))$

$$= \begin{vmatrix} 0.6 & 0.8 & 0.6 & 0.3 \\ 0.6 & 0.8 & 0.6 & 0.3 \\ 0.6 & 0.7 & 0.6 & 0.3 \\ 0.6 & 0.7 & 0.6 & 0.3 \end{vmatrix} \vee \begin{vmatrix} 0.1 & 0.1 & 0.1 & 0.1 \\ 0.2 & 0.2 & 0.2 & 0.2 \\ 0.3 & 0.3 & 0.3 & 0.3 \\ 0.3 & 0.3 & 0.3 & 0.3 \end{vmatrix}$$

$$\begin{vmatrix} 0.6 & 0.8 & 0.6 & 0.3 \\ 0.6 & 0.8 & 0.6 & 0.3 \\ 0.6 & 0.7 & 0.6 & 0.3 \\ 0.6 & 0.7 & 0.6 & 0.3 \end{vmatrix}$$

The value judgement for the given branch service conditions is found by composition, equation 6.2), thus if

A" = [0.7//L+0.6//T+0.7//N+0.6//S]

Then, B" = A"oR

The membership array is given by

$$\mu_V = \text{maxmin} \begin{vmatrix} 0.5 & 0.6 & 0.4 & 0.3 \end{vmatrix} \begin{vmatrix} 0.6 & 0.8 & 0.6 & 0.3 \\ 0.6 & 0.8 & 0.6 & 0.3 \\ 0.6 & 0.7 & 0.6 & 0.3 \\ 0.6 & 0.7 & 0.6 & 0.3 \end{vmatrix}$$

$$= \begin{vmatrix} 0.6 & 0.6 & 0.6 & 0.3 \end{vmatrix}$$

or B" = [0.6//SU+0.6//ST+0.6//LS+0.3//IN]

As might be anticipated, the branch conditions fall below the bench-mark conditions. The value judgement quantifies this and provides a clear impression of the comparison with the bench-mark.

Method study also finds a fruitful field of application in planned maintenance, but the application of fuzzy logic methods here is still at an early stage.

6.6.2 Work measurement

A prime feature of work measurement is that it introduces the notion of time into work study and in broad terms provides a measure of the work content of an activity. The importance of work measurement lies in the ability to provide essential information for the costing of products and resources, operations planning and control, methods of selection and also comparative studies between various locations within an organisation or between organisations.

A key factor is the definition of standard time. This, together with the concept of standardisation of working conditions and equipment enables comparative studies to be made, this is especially important for national for national and international organisations having similar operations in various locations. Standard time is associated with a work element, which is a unit of motion such as: pick up a telephone with the right hand, transfer to the left hand and place on the ear. Elements are aggregated into an activity such as making a telephone call. The activities themselves can be aggregated into operations such as finding the best quotation for a materials supply. Accurate timing of the elements depends upon the choice of breakpoints, which determine where an element begins and ends. Evaluation of observed times also depends upon the concept of the standard rate of working.
British Standard BS 3138:1992 contains a comprehensive review of the terminology used in work measurement and related areas. For the purposes of this text, the following brief comments will suffice.

Standard time

This may be defined as comprising the basic elements which would be the same everywhere for the same activity carried out under all the same operational conditions. To this is added eight different allowances which depend upon local conditions:

i)	Fatigue	v)	Contingent work
ii)	Personal needs	vi)	Contingent delays
iii)	Accessories maintenance	vii)	Unoccupied time
iv)	Relaxation	viii)	Interference

The total work content of an activity is defined as the total basic time of the elements, plus the sum of relaxation, personal needs, tool maintenance and contingent work allowances. Items i),ii),iv),v) and vi) are each expressed as a percentage of the total basic time, but items vii) and viii) are not. Item iii) is expressed as a percentage of the accessories utilisation time. Total allowances would typically range from 5% to 20% of the total basic time.

Elements

For common manufacturing activities involving manual tasks, the elements are usually in the range of about 5 to 30 seconds. For intensive studies microelements in the range of 0.01 to 5 seconds captured on film may be used. On a large scale, for example in the construction industry, elements may range in time from 30 to 1000 seconds. Beyond that, for example in cargo transportation, the elements may range up to weeks, Elements are determined by consensus expert opinion of those involved in the application.

Standard rate of working

This is a subjective benchmark; it means the normal rate of working that a typical worker would achieve under ideal conditions without a break, graded as 100. The standard rate of working will be affected by the ease or difficulty of performing a work element. The ease of operating a parcel delivery service under congested traffic conditions is different from that of operating the same service under light traffic conditions. This effect is known as variance. For example, if and observed time is 30 minutes and the element is rated at 110, then the basic time is given by

$$BT = 30*110/100 = 33 \text{ minutes}$$

In practice it is difficult to ascribe an exact number to the rating and more natural to postulate a fuzzy set representation, giving a more realistic impression of the state of knowledge. Likewise, the total basic time of work elements would be expected to carry an acceptable error and can be effectively represented by a fuzzy set.

If a real or synthetic (from a previous motion and time study) data are available, they may be used as benchmarks, the a relational array between rating and total basic time, or between other system parameters, may be formed. This enables predictions to be made for planning information purposes.

Suppose that two fuzzy system parameters are to be related, say A and B. Then if a logical proposition of the form: IF A THEN B may be used, this provides the following identity

$$\text{IF A THEN B} = (A\times B)\cup(A'\times Y) = R \qquad 6.12)$$

where A is defined on the universe X and B is defined on the universe Y.

In membership function form the relationship is expressed by

$$\mu_R(x,y) = \vee((\mu_A(x)\wedge\mu_B(y)),((1-\mu_A(x)\wedge 1))) \qquad 6.13)$$

Equations 6.12) and 6.13) define the relational array R.

The predictive function is obtained by composition of a postulated set in X, say A", with the relational array R, thus

$$B'' = A''\text{o}R \qquad 6.14)$$

where B" is the predicted set in Y.

If pairs of benchmark parameters, A_i and B_i, (i = 1......n) are available, then the relational array may be extended in form by the union of the n sub-arrays

Let $R_i = (A_i \times B_i)(A_i' \times Y)$ 6.15)

where A_i' is the complement of A_i

and $B_i'' = A''oR_i$ 6.16)

then $B'' = B_1'' \cup B_2'' \cup .... B_n''$ 6.17)

$= A''o(R_1 \cup R_2 \cup .... R_n)$ 6.18)

thus $B'' = A''oR$ 6.19)

where $R = R_1 \cup R_2 \cup .... R_n$ 6.20)

Example 6.6

As a result of prior experience of delivering goods for customers, a transport company has developed a data base with pairs of profiles of driver-condition ratings (P) and average journey speeds (M)

Rating(P)	**Av. Journey Speed (M) m.p.h.**
LP = [0//60+1//60+0.5//70+0//80]	LM = [0//30+1//30+0.5//35+0//40]
AP = [0.5//70+1//80+1//100+0.5//105]	AM = [0//30+1//30+0.5//35+0//40]
HP = [0//100+0.5//105+1//110+0//110]	HM = [0//40+0.5//45+1//50+0//50]

Evaluate the average journey speed for a new route with an estimated driver-condition rating given by

P" = [0//50+0.5//60+1//70+0.7//80+0.5//100]

Solution

The relationship array for the LP-lm pair is first found using equations 6.12) and 6.13)

$$LP \times LM = \min \begin{vmatrix} 0 \\ 1 \\ 0.5 \\ 1 \end{vmatrix} \begin{vmatrix} 0 & 1 & 0.5 & 0 \end{vmatrix} = \begin{vmatrix} 0 & 0 & 0 & 0 \\ 0 & 1 & 0.5 & 0 \\ 0.5 & 0.5 & 0.5 & 0.5 \\ 0 & 0 & 0 & 0 \end{vmatrix}$$

$$LP' \times Y = \min \begin{vmatrix} 1 \\ 0 \\ 0.5 \\ 0 \end{vmatrix} \begin{vmatrix} 1 & 1 & 1 & 1 \end{vmatrix} = \begin{vmatrix} 1 & 1 & 1 & 1 \\ 0 & 0 & 0 & 0 \\ 0.5 & 0.5 & 0.5 & 0.5 \\ 0 & 0 & 0 & 0 \end{vmatrix}$$

Therefore, according to equation 6.15)

$R_i = (LP \times LM) \cup (LP' \times Y)$

Thus,

R_1 =

	30	30	35	40
60	1	1	1	1
60	0	1	0.5	0
70	0.5	0.5	0.5	0.5
80	0	1	0.5	0

In the same way, R_2 and R_3 are constructed from (AO,AM) and (HP,HM) respectively, with the following results:

R_2 =

	30	35	40	50
70	0.5	0.5	0.5	0.5
80	0	0.5	1	0
100	0	0.5	1	0
105	0.5	0.5	0.5	0.5

R_3 =

	40	45	50	50
100	1	1	1	1
105	0.5	0.5	0.5	0.5
110	0	0.5	1	0
110	1	1	1	1

Thus, from equation 6.20)

$$R = R_1 \cup R_2 \cup R_3$$

Hence,

	30	30	35	40	45	50	50
60	1	1	1	1			
60	0	1	0.5	0			
70	0.5	0.5	0.5	1	.	0.5	
80	0	1	0.5	1	.	0	
100	.	0	0.5	1	1	1	1
105	.	0.5	0.5	0.5	0.5	0.5	0.5
110	.	.	.	0	0.5	1	0
110	.	.	.	1	1	1	1

The work rating test profile is given as

P" = [0//50+0.5//60+1//70+0.7//80+0.5//100]

Reforming this to match the relationship array (R) rating scale, then

P" = [0//60+0.5//60+1//70+0.7//80+0.5//100+0//105+0//110+0//110]

Now performing the composition as indicated in equation 6.14) to find the equivalent average journey speed (M") to the work rating P"

$$M" = P" \circ R$$

The following speed profile is obtained

M" = [0.5//30+0.7//30+0.5//35+1//40+0.5//45+0.5//50]

This profile is illustrated in Figure Ex 6.4.

Defuzzifying the speed profile in Figure Ex. 6.4. by the centroid method gives a discrete speed value of 39.6 m.p.h.

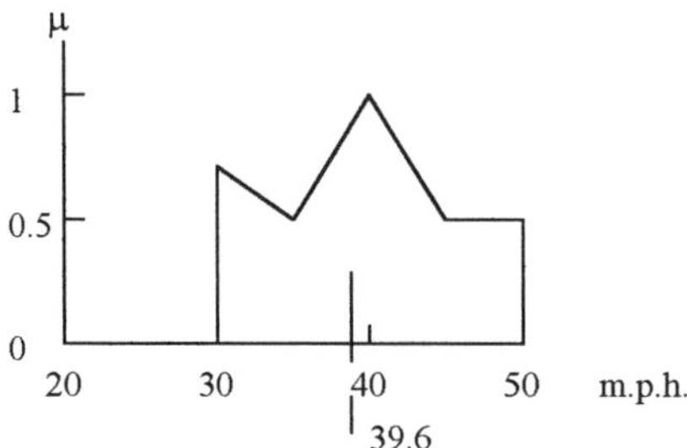

Figure Ex. 6.4 Fuzzy Speed Profile M"

Predetermined motions time study (PMTS)

In the operations planning stage the actual operations motions cannot be observed. It may also be the case that the number of cycles of the operation are too few to make an actual time study practical or economical, for this and various other reasons alternatives are used. For example, it may be more economical to look up a table to determine the standard time that it would take for a person to walk fifteen steps rather than to carry out an actual time study exercise as part of an operations analysis. The use of standard tables summarising PMTS data also tends to average out observation or rating errors of different observers.

If the work study analysis has a PMTS data table available then the proposed or real activity may be synthesised by a logical union of the elements to yield the operation total basic time, and after the addition of allowances, the standard time. There is computer software, including graphics, to facilitate the procedure. Care has to be taken that the elements are infact independent for the process to be valid.

In a fuzzy logic system there is no need to define sharp boundaries, fuzzy sets can naturally accommodate a degree of uncertainty and it will be apparent that elements often blend into each other and so precise boundaries are the difficult to define. For example, if there are fuzzy elements: A,B and C defined on the universe X (time), then the basic times of the elements could be represented by

$$T_A = [\mu_{A1}//x_{A1}+\mu_{A2}//x_{A2}+......\mu_{Am}//x_{Am}]$$
$$T_B = [\mu_{B1}//x_{B1}+\mu_{B2}//x_{B2}+.........\mu_{Bn}//x_{Bn}]$$
$$T_C = [\mu_{C1}//x_{C1}+\mu_{C2}//x_{C2}+.........\mu_{Cp}//x_{Cp}]$$

The total basic time of the elements is the union of the separate elements

$$T = T_A \cup T_B \cup T_C \tag{6.21)}$$

In the FL representation of elements, the tendency for the correct element to be B rather than A, for example, depends upon the cross-over point between the fuzzy sets A and B where they overlap. In conventional work study language the breakpoint always marks the end of element A and the beginning of element B, in the style of CL. If there is no cross-over point in the FL representation of A and B, it means that the activity is discontinuous. At the other extreme, if the membership value at the cross-over point is close to unity, it means that the elements A and B are effectively indistinguishable and may therefore be viewed as one.

The total basic time of several elements in the FL description is illustrated in Figure 6.5.

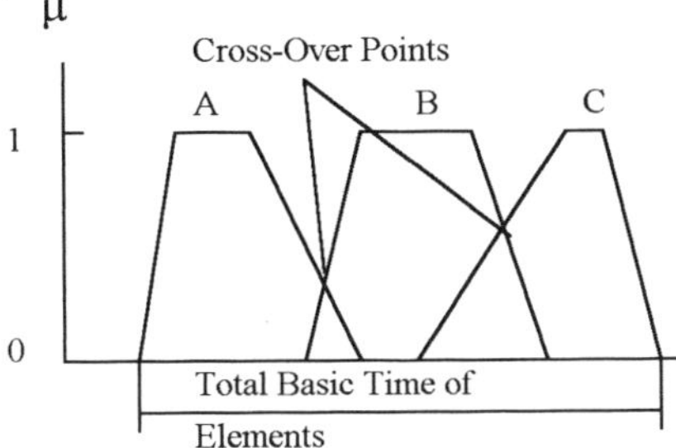

Figure 6.5 Total Basic Time of Several Elements

If the intersection of the fuzzy set with the horizontal (universe of discourse) axis is unclear it is necessary to chose a small value of the membership function and use the intersections with this axis to calculate the total basic time.

Example 6.7

A work measurement conducted on a certain activity produced fuzzy sets for four elements A,B,C and D as follows, where the elements are expressed in minutes

$$T_A = [0//4.58+0.6//4.84+1//5.09+0.6//5.34+0//5.60]$$
$$T_B = [0//5.70+0.5//5.94+1//6.19+0.7//6.43+0//6.68]$$
$$T_C = [0//4.27+0.2//4.45+1//4.62+0.4//5.00+0//5.31]$$
$$T_D = [0//6.90+0.8//7.11+1//7.30+0.8//7.52+0//7.73]$$

Determine the total basic time of the elements comprising the activity.

Solution

A graphical solution is adequate for this problem since the accuracy depends upon the estimates of the time keeper. The fuzzy elements are illustrated in Figure Ex. 6.6. B and D do not have any overlap with any other element, this occurs with fragmented activities. Elements A and C have an overlap with a cross-over point at 4.75 minutes.

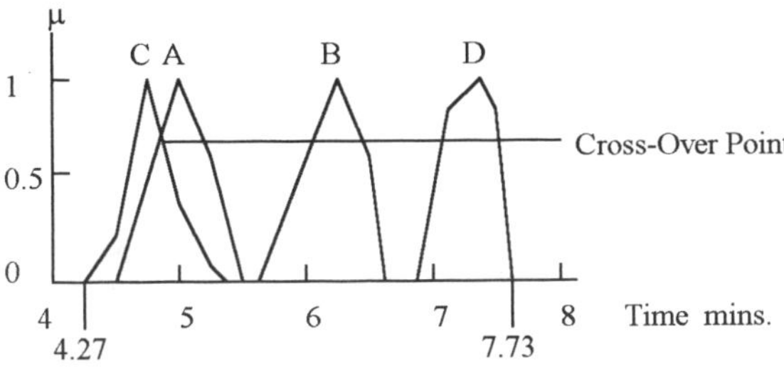

Figure Ex. 6.6 Basic Time of Elements (7.73-4.27)

By inspection of Figure Ex. 6.6, the total basic time of the four elements is 7.73-4.27 = 3.46 minutes.

The total time that a product occupies in a given process is known as the product cycle time. It is the sum of all the activity times within that process. This is the basis for estimating the number of similar parallel processes required to achieve a given output rate of goods or services. It also sets the time for materials requirements planning.

6.7 Process planning

In a production system for goods or services, the entire process characteristics must be known to enable the best use of resources to be achieved which are compatible with the objectives of the organisation, (which may or may not be that of maximising profits, depending upon the nature of the organisation). In the manufacturing and utilities sectors the method of achieving this is known as process or production planning and it encompasses all the value adding conversions and ancillary functions from the input of resources of the system, to the final delivery of finished products or services.

The design of the processes to be used in such a system proceeds concurrently with the development of the goods or services in modern practice, because of the universal pressure for reducing the concept-to-delivery times. The management team must own a knowledge bank about possible technological and other choices, or it has to make good any shortfall by buying such knowledge. The knowledge must also be embodied in accessible form, such as data banks in various formats, including fuzzy logic rule-bases.

Factors affecting judgements in process planning include the total output demand pattern (which affects capital resources investment), rate of demand, available in-house knowledge and intelligence (human and artificial intelligence) and the perceived benefits which will accrue from the proposed operations. These will be balanced against the total resources investment required to establish and maintain the operations, and also the opportunity cost.

A key factor lies in choosing the type and specification of the technology that will enable the transformations required to be made, support the total output and sustain the rate of demand pattern. In the industrial sector there will be data knowledge banks for the unit operations in manufacturing and processing to guide planning. Much of this knowledge implicitly uses a CL base, which can contribute to difficulties in its application. FL is able to provide a more realistic portrayal of the possibilities and to capture the change over from one type of operation to another as the product parameters change.

Example 6.8

There is a demand for a moulded component in brass, which is required at the rate of 45 pieces per day. The geometry of the moulding is mainly hollow cylindrical, but it also has a hollow non-cylindrical shape occupying about one third of the volume.

The manufacturing company planning which is planning the possible production of the piece has its moulding knowledge embedded in the following rule-base:

Moulding Geometry	**Production Rate** (Mouldings per Day). 20	40	60	80
S	B	B	C	C
M	A	B	B	C
C	A	A	A	A

The geometric classification is as follows:
S = solid or hollow cylindrical shape with one or more outside diameters
M = solid or hollow non-cylindrical shape
C = flanged and complex non-cylindrical shape

The recommended moulding methods are:
A = investment casting
B = gravity die casting
C = plaster moulding
D = shell moulding

Determine the appropriate moulding method for the proposed production.

Solution

The production rate table may be framed in an equivalent fuzzy rule-base form as follows:

G	**P** LO	LM	HM	HI
S	B	B	C	C
M	A	B	B	C
C	A	A	A	A

where P = production rate.
G = geometry.
LO = low.
LM = low medium.
HM = high medium.
HI = high.

The corresponding partitioning of the P,G and M universes are shown in Figure Ex. 6.7.

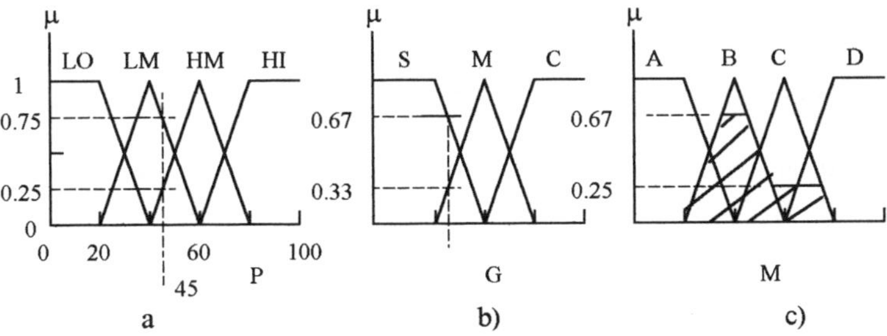

Figure Ex. 6.7 Partitioning of the Universes of Discourse for the Choice of Moulding Method
a) Production Rate b) Class of Geometry c) Class of Moulding Method

In this problem there are two antecedents (input parameters), namely, the production rate and the class of geometry, therefore a logical proposition of the form: IF A AND B THEN C is appropriate. Using the required production rate of 45 pieces per day, the membership values may be obtained from Figure Ex. 6.7 a). as 0.75 LM and 0.25 HM. The information about the moulding geometry implies that it is 67% hollow cylindrical and 33% hollow non-cylindrical, this is illustrated in Figure Ex. 6.7 b), giving 0.67S and 0.33M. Partial conclusions may now be drawn using the above form of logic proposition

IF P	AND G	THEN M	MIN.	CONCLUSION
LM	S	B	0.75,0.67	0.67 B
LM	M	B	0.75,0.33	0.33 B
HM	S	C	0.25,0.67	0.25 C
HM	M	B	0.25,0.33	0.25B

Partial conclusions are included in Figure Ex. 6.7 c). The overall conclusion is the union of the partial conclusions. The choice clearly falls into the gravity die cast class (B) of production methods. The ratio of the B to C areas in Figure Ex. 6.7 c) is about 3.58:1. Using the original table of information given in the problem, the out come is not so clear cut, thus showing the advantage of the FL method.

The treatment outlined above may be extended to different levels of generality, for example to choose between subcontracting (or out-sourcing) fabrication or moulding. The appropriate production technology also depends to a certain extent upon the size of the product; all technologies are not applicable to all product sizes, for example, pressure die-casting is limited to smaller sizes of product. For large complex units which comprise sub-assemblies, there are many possible combinations of subcontracting and in-house manufacture and assembly. For service providers analogous considerations apply; financial services may offer assemblies of financial products to different market segments. Larger organisations are able to offer their own products as well as assembled bought-in products. At the present time, rule-bases are quite specific, though in time it may be expected that some generality will be achieved in a number of industrial and service sectors.

6.8 Inventory control

It is obvious that the existence of inventory (stock) within an organisation represents both frozen capital and loss of revenue by inventory maintenance cost and opportunity cost, for example, maintaining patients in hospital. Inventory reduction has therefore been the focus of much attention, seen as a way of increasing operational and financial efficiency. Although much of the published work in this area has been concerned with manufacturing, the concepts have wider applications.

There are at present two widely used approaches to inventory control:
i) Material requirements planning (MRP)
ii) Just-in-Time (JIT) scheduling

MRP provides management information about the scheduling of goods and services that need to be requisitioned throughout the entire organisation, consistent with the perceived production plan. It reflects the interdependency of upstream and downstream processes in a balanced production system. MRP is suitable for both batch and continuous flow systems. In both types of system, the groups of common specification items are entered into the processing system in cyclic succession, in accordance with the production schedule. The concept of MRP has been extended by making the information used the basis for general operational and financial planning (including sales, cash flow and purchasing). This extension is called materials resources planning (MRP II).

The alternative so-called JIT system aims at a continuous flow of materials through the processing system from the delivery of inputs to the exit of finished goods. In the ideal case the only inventory is work in progress, This ideal can only be approached in the limited circumstances of:

i) Continuous mass production
ii) Uniform products
iii) Industrialised regions where continuity of external supplies is assured
iv) Unrestricted inputs
v) Continuous demand

Within the organisation a kanban (card) system is used to signal a demand for the supply of more upstream products to reduce inventories of semi-finished products. It is therefore considered to be a pull system, with production being pulled through the processing system from the exit end. In practice, the JIT system is approximated by having small batch sizes, with the smallest possible change-over times for tooling set-ups, and also low inventories of inputs, semi-finished and finished goods. In reality an organisation has to well adapted to the exigencies of its situation, which can be quite restrictive in less industrialised regions. In this case considerations of different types of stock can become particularly significant. Usually, three types of stock are identified with their uses as follows:

i) Active - Regular demands which are known.
ii) Buffer - Variations to regular demands.
iii) Strategic - Physical, against shortages and emergencies. Financial, against anticipated currency movements.

As a first step in inventory reduction, a Pareto analysis can provide an effective basis for reducing the financial penalty of stock holding. This is a general concept and may be applied when a relatively small fraction of the population accounts for a large fraction of a particular population characteristic. The applicability of a Pareto analysis to a manufacturing process system may be seen from the fact that the processes are value adding operations in different proportions. As the material moves through the system the unit value increases as illustrated in Figure 6.6.

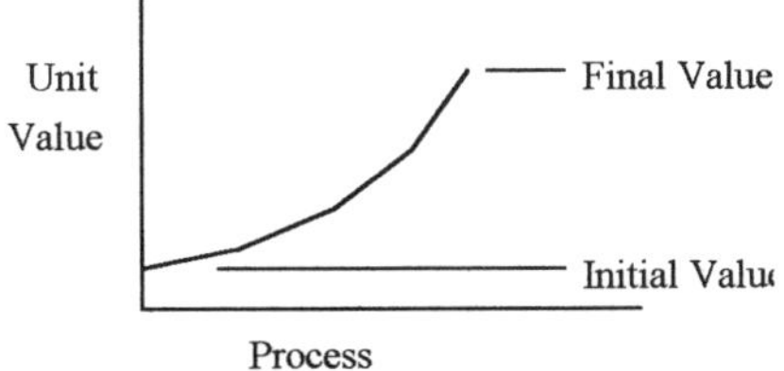

Figure 6.6 Pareto Diagram for Inventory Unit Value

It is illustrated in Figure 6.6 that product unit value increases as it passes through the production system at an increasing rate and that the last fraction of the output contains much more value than the initial fraction. Therefore inventory reduction should proceed upstream from the exit end to be most effective. The unit value should include all costs up to that point, including overheads.

6.9 Batch processing

The material resources plan is compiled after the master production schedule is created. This in turn depends upon the production plan and the overall business plan. At the master production schedule stage the optimum production batch size is estimated and also the production cycle is chosen. Various idealised cyclic models may be postulated with different levels of complexity to represent reality more closely. Invariably these are based upon deterministic treatments, though the underlying information is in fact fuzzy. Some of the better known models are outlined below, followed by their recasting in FL terms.

Simple model

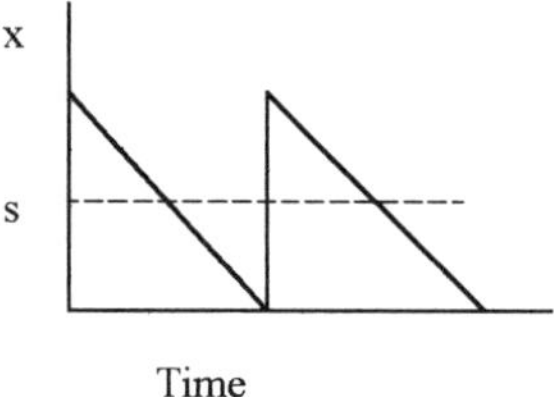

Figure 6.7 Simple Batch Processing Model

The simplest deterministic model of batch processing is illustrated in Figure 6.7, showing a constant demand rate and instantaneous production. The conditions for the minimum total cost per unit time may be found for this idealised model as follows:

Let x = volume of product per batch.
s = average inventory level.
m = sum of storage and opportunity cost per product per unit time.
n/x = process set-up cost.

The total cost per unit time is given by

$$c(x) = ms+n/x \qquad 6.22)$$

$$= mx/2+n/x \quad \text{(since } s = x/2\text{)} \qquad 6.23)$$

This expression is a maximum or minimum on dc(x)/dx = 0. That is when

$$x_0 = (2m/n)^{1/2} \qquad 6.24)$$

It is easily seen that d(c)/dx is +ve for all real values of x, therefore equation 6.23) does give a minimum, its value is

$$c_0 = 2(2mn)^{1/2} \qquad 6.25)$$

Let p be the product demand rate, then the process cycle time is x/p. The average inventory level is s (constant).

Although this discussion is phrased in terms of production, it is equally applicable to the purchasing of input resources. The lead time for reordering may then be governed by an inventory level, or alternatively, by a fixed interval order cycle with variable order level.

Finite processing rate

A more realistic model is obtained by allowing for a finite production rate, say p. The inventory pattern is then modified as illustrated in Figure 6.8. The average inventory is now less than in the previous case.

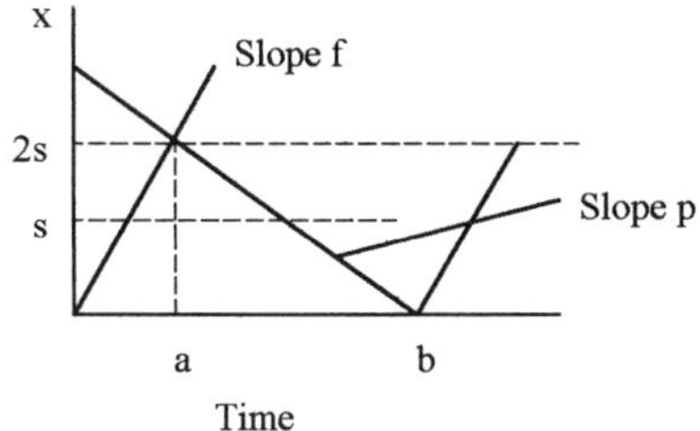

Figure 6.8 Idealised Batch Processing with a Finite Production Rate

By inspection of Figure 6.8

$$2s/x = 1\text{-}a/b \qquad 6.26)$$

Also, $$x/2s = pb/fa \qquad 6.27)$$

Combining equations 6.26) and 6.27) gives

$$s/x = 1/(2(1+p/f)) \qquad 6.28)$$

or, $$s = qx \qquad 6.29)$$

where $q = 1/(2(1+p/f))$

Combining equations 6.22) and 6.29) results in a deterministic expression

$$c(x) = mqx+n/x \qquad 6.30)$$

Differentiating this expression to find the minimum yields the inventory level for minimum cost

$$x_1 = n/mq \qquad 6.31)$$

Instead of equation 6.25), the minimum cost is now given by

$$c_1 = 2(mnq)^{1/2} \qquad 6.32)$$

The process idle time is (b-a). After some algebraic manipulation this is found to be

$$t = b\text{-}a = xf/p(f+p) \qquad 6.33)$$

6.10 Cyclic batches

Slack time in an adaptable process means that it has surplus capacity. Clearly, the system utilisation can be increased by setting it up to process different work in the slack time available. This case exists wherever the production rate is higher than the demand rate. In the case of it being less, a parallel process is needed, then the pair may present surplus capacity.

Consider two products each using the same process station; the demand-production cycles are modelled in Figure 6.9.

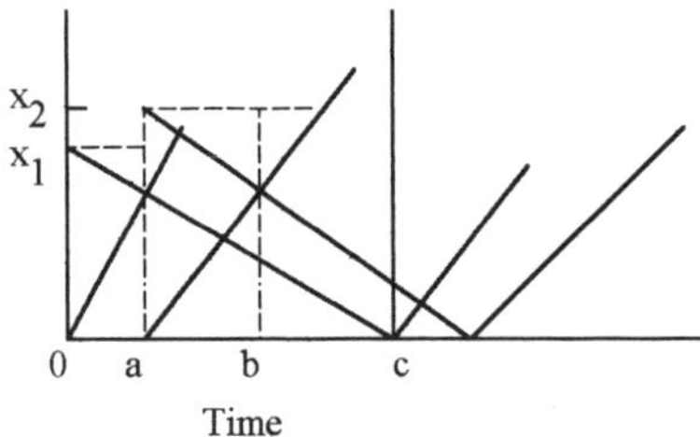

Figure 6.9 Batch Processing with a Finite Production Rate

Considering Figure 6.9, the first batch processing commences at zero time and the first batch of products, x_1, is completed at time a. If the change-over time is negligible, the second batch of products, x_2, is commenced and the processing is completed at time b, before the stock of the initial batch is exhausted at time c. At time c, the second batch of the product, x, can be commenced. The process idle time is now less (c-b) than is the case for a single product (c-a). The demand for the second type of product needs to be sufficiently low for this to be completed before the next change-over is due. Following this through to more products can yield interesting and complex production patterns. Such a development will not be persued here, it is of more immediate interest to consider fuzzy cases.

Although the discussion is phrased in terms of production, the same concepts may also be applied to the purchasing of input resources.

6.11 Fuzzy batch processing

Fuzzy categories may arise in a number of ways in processing systems. Consider a model system which is the fuzzy equivalent of equation 6.30), in which the coefficients m and n are not known exactly. This means that the opportunity and change-over costs can only be approximately estimated, but may be specified within a range. The deterministic form of the preferred batch size is given by equation 6.31).

Let n,m and q become fuzzy sets N and H where H represents the fuzzy set of the product mq. For the following discussion, let N and H be defined by three-term fuzzy sets, viz

$$H = [\mu_{11}//h_{11}+\mu_{12}//h_{12}+\mu_{13}//h_{13}] \qquad 6.34)$$

and

$$N = [\mu_{21}//n_{21}+\mu_{22}//n_{22}+\mu_{23}//n_{23}] \qquad 6.35)$$

The fuzzy set for the preferred batch size, X, is a function of H and N, thus

$$X = \max(\min(\mu_{11},\mu_{21}))//(h_{11}^{-1}n_{21})+\max(\min(\mu_{11},\mu_{22}))//(h_{11}^{-1}n_{22})+\ldots$$
$$\ldots+\max(\min(\mu_{13},\mu_{23}))//(h_{13}^{-1}n_{23}) \qquad 6.36)$$

The minimum cost, the production time and also the process idle time now become fuzzy sets.

Example 6.9

It is estimated that the opportunity and overhead costs of a product are 0.2% per week of the processing cost of £5.00 per product unit. The process set-up cost is £2.00. The product demand is 80 units per week and the planned production rate is 200 per week. The significant uncertainty of the opportunity costs and set-up costs is considered to be ±20%.

Estimate the preferred batch size, the minimum cost, the maximum inventory of the product and also the process idle time.

Solution

From equations 6.28) and 6.29)

$q = 1/(2(1+p/f))$ where p = 80 and f = 200

hence, $q = 0.3571$

Now $m = 5*0.2/100 = 0.01$ Therefore, $mq = 0.3571*0.01 = 0.003571$

and $n = 2*80 = 160$

Expressing mq and n as fuzzy sets, each with a ±20% range, then

$$mq = [0.5//0.003+1//0.0036+0.5//0.00421]$$

and
$$n = [0.5+128+1//160+0.5//192]$$

The preferred batch size fuzzy set, X, is given by equation 6.36). The element values are

$h_{11}^{-1} = 333.3$ $\quad n_{21} = 128$
$h_{12}^{-1} = 277.8$ $\quad n_{22} = 160$
$h_{13}^{-1} = 238.1$ $\quad n_{23} = 192$

The elements of X, rounded off to the nearest integer and the corresponding membership values are tabulated below.

	h_{11}^{-1}	h_{12}^{-1}	h_{13}^{-1}		μ_{11}	μ_{12}	μ_{13}
n_{21}	207	**189**	**175**	μ_{21}	0.5	**1.0**	**0.5**
n_{22}	**231**	211	195	μ_{22}	**1.0**	1.0	1.0
n_{23}	**253**	**231**	214	μ_{23}	**0.5**	**1.0**	0.5

The principal values of the elements and membership values of the X fuzzy set are shown in bold type in the above tabulation. The X fuzzy set is illustrated in Figure Ex. 6.8.

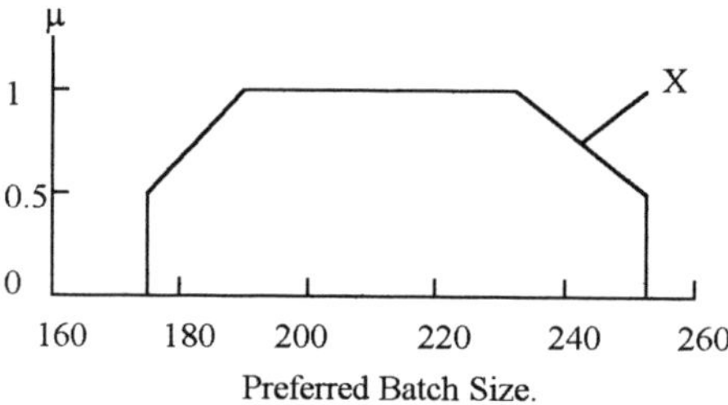

Figure Ex. 6.8 Preferred Batch Size X

Defuzzifying the above diagram by the centroid method indicates a preferred batch size of 213 to the nearest integer.

The minimum cost may be evaluated with the aid of equation 6.32). The results are tabulated below.

$h_{11} = 0.003$ $\quad n_{21} = 128$
$h_{12} = 0.0036$ $\quad n_{22} = 160$
$h_{13} = 0.0042$ $\quad n_{23} = 192$

h	h	h	
n	**1.24**	**1.36**	1.47
n	1.39	1.52	1.64
n	1.52	**1.66**	**1.80**

Principal values of the C fuzzy set are shown in bold type in the above tabulation. The corresponding membership values are as given above for the X fuzzy set. The C fuzzy set is illustrated in Figure Ex. 6.9.

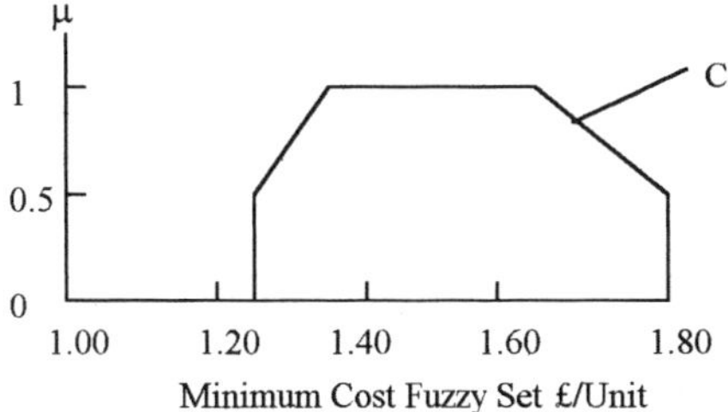

Figure Ex. 6.9 Minimum Cost Fuzzy Set C

From equation 6.29) the maximum inventory level (2s) is 2qx, where

$$2q = 1/(1+p/f) = 1/(1+80/200) = 0.3571$$

The X element table given above may be factored by 0.3571 to produce the following maximum inventory level (2S) results

	h_{11}	h_{12}	h_{13}
n_{21}	148	**134**	**124**
n_{22}	**164**	150	140
n_{23}	**180**	**164**	152

The illustration will obviously be a scaled diagram of Figure Ex. 6.8.

The following results for the idle time (T) are also derived from the X element table above

	h	h	h
n	1.55	**1.42**	**1.31**
n	**2.06**	1.88	1.74
n	**2.26**	**2.06**	1.91

For both the 2S table and the T table elements given above, the corresponding membership values are the same as for the X table elements. The associated fuzzy set diagrams are also similar to that illustrated in Figure Ex. 6.8.

Within the uncertainty of the data given in the problem, the fuzzy diagrams provide a clear impression of the range of possibilities of the conclusions.

6.12 GANTT charts

This type of chart is a scheduling tool and it is a convenient and frequently used method of illustrating the phasing of associated activities during the planning and operational stages of a project, made by compiling a horizontal bar chart showing activity durations by parallel bars. A typical example is shown below in Figure 6.10.

In this project there are 10 activities spread over a period of 24 weeks. The chart shows the phasing of each of the activities, when they must start and end. The activities A,B,C.......K are placed on the vertical axis and the duration on the horizontal axis. The duration of each activity is given at the end of the appropriate activity bar. The vertical lines are links to successive activities, showing the latest time that an activity can end and the earliest that the successive activity can start.

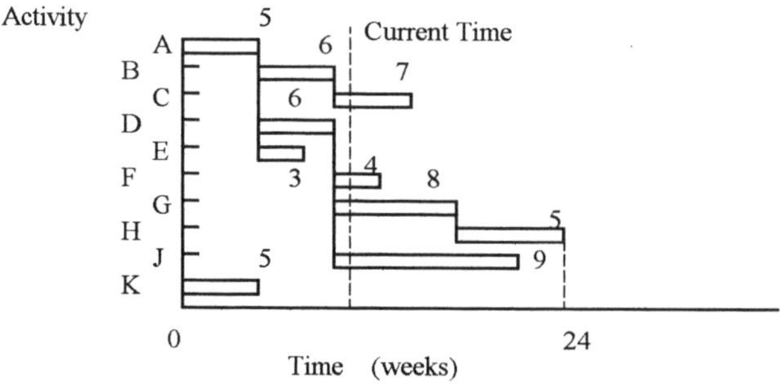

Figure 6.10 A Typical GANTT Chart

It is immediately apparent that the earliest that the group of activities can end is controlled by the most extreme right-hand bar, which is activity H in this case. The end point is linked to the start at zero time by the activities A,D,G,H, this is called the critical path line and the minimum duration (24 weeks) is given by the sum of the critical pathline durations. The current time is shown on the diagram and markers may be needed to show the current position of the activities. It may be noted that the non-critical activities have spare time, called float.

The appearance of the GANTT chart implies that all activities commence and terminate instantaneously. This is an idealised model and in reality there is often unaccounted for time at both the inception and termination of activities and certainly for the whole project. This may, for example, involve briefings or preparatory and post-operations reworking of problems and sub-standard work, successive activities may therefore overlap in the non-ideal case. It is obvious that the normal GANTT chart is based upon CL concepts; an activity has two membership values (0,1). A more articulate form of the chart, in which the activity level in the interval 0 to 1 corresponds with the membership value of the fuzzy set, would be three dimensional as illustrated in Figure 6.11.

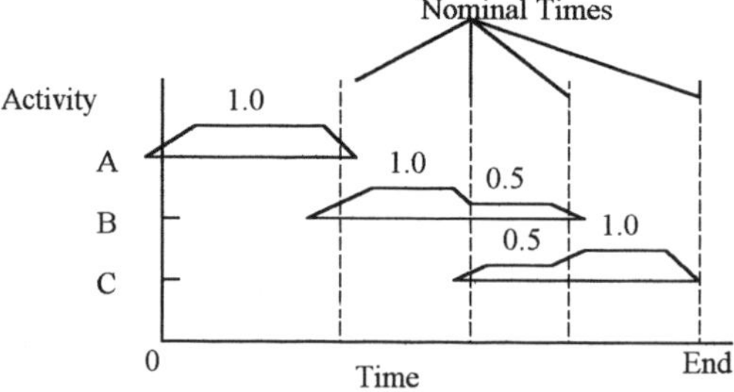

Figure 6.11 Three-Dimensional Fuzzy GANTT Chart

In Figure 6.11, the inception and termination times of both the whole project and those of the individual activities are nominal times. Although only activity levels of 0, 0.5 and 1.0 are shown on the chart, it will be understood that generally any level in the range, 0 to 1 is admissible on this type of chart. The chart allows for both the winding-up and winding-down of activities and also allows for the overlap of activities functioning at less than full strength. Thus if due to some contingency, such as demands of other work, staff sickness or materials shortage activity B was run at a level of 0.5 strength for a period, it may be possible to start a reduced level of activity C to run in parallel for a period.

An alternative form of presentation of the same information is by the network diagram based on CL, which also permits a critical path analysis, but not a FL representation. The GANTT chart is an analogue representation that is easy to read and also widely used in industry.

6.13 Process and product quality

High volume production of quality goods and services is a phenomenon which has its roots in the twentieth century manufacturing and processing industrial sector. Prior to this, the concept of quality was usually only associated with the creation of specimen craftsman and artistic goods for the satisfaction of high nett worth consumers. In the early stages of high volume quality, inspection of the final product for conformance to standard was the normal custom, with reworking or scrapping of substandard items. At the end of the twentieth century, this practice is still common. There is however a trend towards viewing inspection of products as only a diagnostic tool for process control, or even as being unnecessary. The emphasis on quality of process upstream as far back as the supplier inputs organisation and also laterally into the associated managerial and administrative processes is gaining credence.

The concept that quality is automatically conferred on products, whether they are goods or services, by high and consistent process quality is now receiving global recognition and has a strong advocate in the form of the ISO 9000 series of standards. The series comprises ISO 9000 to ISO 9004 (inclusive) and is a set of generic standards covering; definitions and concepts (ISO 9000), quality assurance in design, development, production, installation and servicing (ISO 9001), production and installation (ISO 9002), test and inspection (ISO 9003) and also quality management and quality system elements (ISO 9004).

Ideally, the quality of goods or services is defined by the consumers. In practice they may or may not be able to formulate the concept in exact numerical or linguistic terms and so to reconcile what is technically and economically feasible with aspirations. In many important cases there are national and international standards for goods, especially where matters of health or safety are concerned. It is anticipated that FL methods will find increasing application in this area, particularly where services and other matters that are difficult to quantify are concerned.

Benchmarking, standards and feedback are commonly used tools for the management of quality. Feedback is the same in concept as that which is familiar in automatic control, and in this case means the auditing of goods or services to correct and control the creative processes. Quality control is an important part of quality management.

6.14 Benchmarking for quality

In the continuous improvement of quality philosophy, benchmarking of peer groups means the comparison of an organisation's functional parameters with those of the most successful peer organisations. At its most comprehensive, it covers all the salient features of the technical, financial, managerial and administrative structures and operations, extending also to the input suppliers.

One of the benefits attributed to benchmarking is that it provides a stimulus and a guide to the transfer of improved methods of operations that are cost effective for the recipient. The result of benchmarking is not only the identification of performance deficiencies, but also a means of their reduction or elimination. But the knowledge transfer does require a suitable receptor organisation, this fact is sometimes overlooked. The methods of benchmarking are of course applicable to public services as well as commercial organisations. In the UK the comparisons are often known as league tables for schools, hospitals and other public services, though these are not comprehensive benchmarking studies, generally being too restrictive.

Benchmarking may be carried out at various organisational levels. For example, at the strategic level the following parameters might be considered:

- Human resources
- Financial resources
- Business development plans
- ISO (or equivalent) certification
- Market share
- Sustainability
- Financial indicators

Functional benchmarking might include the following parameters:

- Technical and general service support
- Product reliability
- Price/performance of the product
- Product design rating
- Research and development activity

Benchmarking procedure can be framed in computer spreadsheet form, but conclusions are not so easy to formulate on this basis. An itemised league table is a popular form for publishing benchmarking results, such as the following comparison of Education Authority school results:

	Local Education Authority		
Subject	A	B	C.......................
English	71.1	66.2	61.0..................
Maths	65.2	56.8	54.7..................
Science	73.2	67.7	67.6..................
Aggregate	209.5	190.7	183.3................

The aggregate scale is used as a basis of comparison between the several organisations, the highest of which may be regarded as the reference model. The league tables thus presented are devoid of any consumer reaction and is not comprehensive in terms of quality. Benchmarking in the FL method involves a relational array containing not only subject data but also an evaluation.

The FL method is to create a reference array, which is a knowledge base, not just a data base, since it includes data and evaluation. Then by composition of a given case data with the reference array, to form a conclusion. The reference array is found using equation 6.1), repeated here for convenience

$$R = (X \times Y) \cup (X' \times 1) \qquad 6.1)$$

Consider a simple single reference system and let the benchmarking factors and weights of a product be agreed by consensus to be

Factor	**Weight**
Quality (Q)	μ_q
Price (P)	μ_p
Features (F)	μ_f
Availability (A)	μ_a
Market share (M)	μ_m

The product profile is then given by a FL set

$$W = [\mu_q//Q+\mu_p//P+\mu_f//F+\mu_a//A+\mu_m//M]$$

An independent evaluation of the product could be set in the following terms

Category	**Rank**
Superior (S)	μ_s
Excellent (E)	μ_e
Good (G)	μ_g
Substandard (S)	μ_u
Reject (R)	μ_r

This represents an overall product quality assessment, given in terms of a FL set by equation 6.31)

$$C = [\mu_s//S+\mu_e//E+\mu_g//G+\mu_u//U+\mu_r//R] \qquad 6.37)$$

The equivalent form of equation 6.1) is given by

$$R = (W \times C) \cup (W' \times 1) \qquad 6.38)$$

This expression provides the benchmark array. For a given case, let the product profile be W', then corresponding with equation 6.2), the evaluation is

$$C' = W' o R \qquad 6.39)$$

Equations 6.3) to 6.6) inclusive also apply.

If there is more than one reference system with arrays: $R_1, R_2 \ldots\ldots\ldots R_n$, then the total reference system is provided by the union of the components, Thus

$$R = R_1 \cup R_2 \cup \ldots\ldots\ldots\ldots R_n \qquad 6.40)$$

Example 6.10

It is required to benchmark a product and it is decided to compare it with a well-known and established product which has been the subject of market research with the following results:

Factor	**Benchmark Value**	**Test Product Value**
Quality (Q)	0.8	0.6
Price (P)	0.4	0.7
Features (F)	0.6	0.5
Durability (D)	0.8	0.8
Market share	0.6	0.3

The value judgement of the benchmark is

Grade	**Rank**
Superior (S)	0.7
Excellent (E)	0.8
Good (G)	0.4
Substandard (U)	0.2
Unusable (R)	0.1

Evaluate the product.

Solution

The reference benchmark has the following factors set

$$W = [0.8//Q+0.4//P+0.6//F+0.8//A+0.6//M]$$

The grade set of the reference benchmark is

$$C = [0.7//S+0.8//E+0.4//G+0.2//U+0.1//R]$$

Therefore,

$$\min(\mu_W, \mu_C) = \begin{vmatrix} 0.8 \\ 0.4 \\ 0.6 \\ 0.8 \\ 0.6 \end{vmatrix} \times |0.7 \quad 0.8 \quad 0.4 \quad 0.2 \quad 0.1| = \begin{vmatrix} 0.7 & 0.8 & 0.4 & 0.2 & 0.1 \\ 0.4 & 0.4 & 0.4 & 0.2 & 0.1 \\ 0.6 & 0.6 & 0.4 & 0.2 & 0.1 \\ 0.7 & 0.8 & 0.4 & 0.2 & 0.1 \\ 0.6 & 0.6 & 0.4 & 0.2 & 0.1 \end{vmatrix}$$

Also,

$$\min((1-\mu_W),1) = \begin{vmatrix} 0.2 \\ 0.6 \\ 0.4 \\ 0.2 \\ 0.4 \end{vmatrix} \times |1 \quad 1 \quad 1 \quad 1 \quad 1| = \begin{vmatrix} 0.2 & 0.2 & 0.2 & 0.2 & 0.2 \\ 0.6 & 0.6 & 0.6 & 0.6 & 0.6 \\ 0.4 & 0.4 & 0.4 & 0.4 & 0.4 \\ 0.2 & 0.2 & 0.2 & 0.2 & 0.2 \\ 0.4 & 0.4 & 0.4 & 0.4 & 0.4 \end{vmatrix}$$

Now, $\mu_R = \max(\min(\mu_W,\mu_C)\min(1-\mu_W),1))$

$$= \begin{vmatrix} 0.7 & 0.8 & 0.4 & 0.2 & 0.1 \\ 0.4 & 0.4 & 0.4 & 0.2 & 0.1 \\ 0.6 & 0.6 & 0.4 & 0.2 & 0.1 \\ 0.7 & 0.8 & 0.4 & 0.2 & 0.1 \\ 0.6 & 0.6 & 0.4 & 0.2 & 0.1 \end{vmatrix} \vee \begin{vmatrix} 0.2 & 0.2 & 0.2 & 0.2 & 0.2 \\ 0.6 & 0.6 & 0.6 & 0.6 & 0.6 \\ 0.5 & 0.4 & 0.4 & 0.4 & 0.4 \\ 0.2 & 0.2 & 0.2 & 0.2 & 0.2 \\ 0.4 & 0.4 & 0.4 & 0.4 & 0.4 \end{vmatrix}$$

that is

$$\mu_R = \begin{vmatrix} 0.7 & 0.8 & 0.4 & 0.2 & 0.2 \\ 0.6 & 0.6 & 0.6 & 0.6 & 0.6 \\ 0.6 & 0.6 & 0.4 & 0.4 & 0.4 \\ 0.7 & 0.8 & 0.4 & 0.2 & 0.2 \\ 0.6 & 0.6 & 0.4 & 0.4 & 0.4 \end{vmatrix}$$

In the trial case, the factors fuzzy set is

W' = [0.6//Q+0.7//P+0.5//F+0.8//A+0.3//M]

Therefore, from equation 6.39), by composition

$$\mu_C = |0.6 \quad 0.7 \quad 0.5 \quad 0.8 \quad 0.3| \begin{vmatrix} 0.7 & 0.8 & 0.4 & 0.2 & 0.2 \\ 0.6 & 0.6 & 0.6 & 0.6 & 0.6 \\ 0.6 & 0.6 & 0.4 & 0.4 & 0.4 \\ 0.7 & 0.8 & 0.4 & 0.2 & 0.2 \\ 0.6 & 0.6 & 0.4 & 0.4 & 0.4 \end{vmatrix}$$

$$= |0.7 \quad 0.8 \quad 0.6 \quad 0.6 \quad 0.6|$$

Hence, the resultant evaluation of the product trial case is

C' = [0.7//S+0.8//E+0.6//G+0.6//U+0.6//R]

The C and C' fuzzy sets are illustrated in Figure Ex. 6.10.

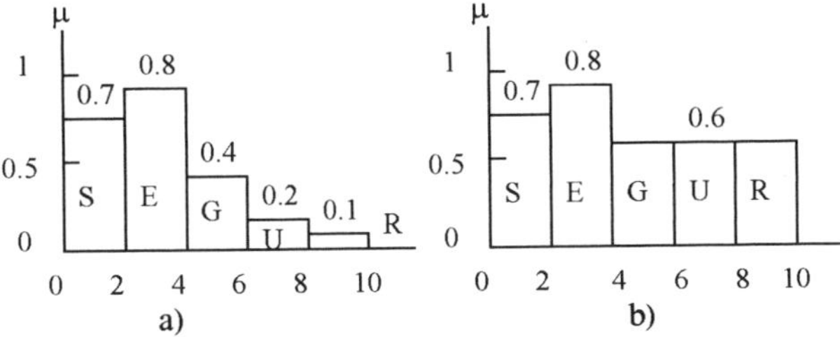

Figure Ex. 6.10 Fuzzy Sets of: a) Benchmark Case b) Test Case

Figure Ex. 6.10 compares the two cases of C and C', and by inspection of the diagrams it may be seen that though the test case and the benchmark cases are equally weighted in the superior and excellent categories, the test case is more heavily weighted than the benchmark in the lower categories. This can be brought more sharply into focus by finding the centroids of the two diagrams. For the benchmark case the centroid value is 3.36, which is almost mid-point excellent, whilst for the test case it is at 4.75, which is almost mid-point good, a lower category than excellent. A higher level of consumer dissatisfaction may therefore be anticipated.

6.15 Quality control

This is a crucial part of quality management, and by means of regulation of the relevant processes, the quality of goods or services is made to conform to the quality agreed with the consumers, assuming that this is technically and economically feasible. To be fully effective, quality control should be an integrated part not only of the product processing and delivery systems, but also of the associated administrative, management and support services. The focus of quality control should be to identify process malfunctions and to implement regulatory actions to rectify these to enable the required product quality to be matched and by continuous improvement, to be exceeded.

The usual quality control diagnostic tools are:

i) Flow charts for process design.
ii) Cause-and-effect (fishbone or Ishikawa) diagrams as illustrated in Figure 6.12.
iii) Histograms for failure frequency occurrence analysis.
iv) Pareto diagrams for failure type ranking.
v) Correlation diagrams for cause-effect association.

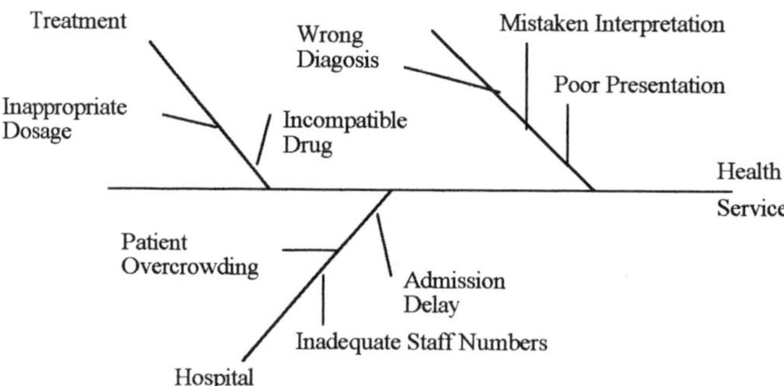

Figure 6.12 Cause-and-Effect Diagram for a Health Service

Each of the causes of quality faults given in Figure 6.12 can be viewed on a FL basis, rather than on a CL (0,1) basis, and be ascribed a membership value. The universe of faults then comprises a number of fuzzy sets which together make up the total evaluation of the Health Service case

6.16 Process quality control

The normal concepts of process control are based upon a process which produces a more or less continuous stream of goods or services. Control is exercised on those parts of the process which affect the quality of the product, quality being expressed as an agreed specification range within an upper specification limit (USL) and a lower specification limit (LSL).

There are two main facets to process analysis, these are: process capability analysis and control charts. They usually assume that the quality factors of interest can be satisfactorily represented by one of the common statistical distributions; normal, binomial or Poisson.

The capability of a process is defined as the range of values to be expected for a process under statistical control (usually with a normal distribution). This is commonly taken as $\pm 3\sigma$ where σ is the standard deviation. If (USL-LSL)/6σ is greater than unity, then the process should produce very few or zero sub-standard products.

Process control by control charts is intended to detect non-statistical variability which may then be identified with specific causes within the process. Regular sampling of the product stream is conducted and the relevant statistical parameters are calculated and scrutinised for: out-of-limit strays, shift of average values, trends (drifts) and cyclic variations. The process is then corrected to eliminate any of the non-statistical variability.

Quality metrics are usually classed as measurement by variables when the relevant parameters are accessible to direct measurement, or measurement by attributes when the parameters are not measurable. In the latter case, the product is then examined for conformance to specification or standard and the quantity of sub-standard or non-conforming product is recorded. The USL and LSL then refer to the quantity of non-conforming samples. Further details of sampling parameters, such as sample size and frequency, may be found in most texts on statistical control methods.

Fuzzy process quality control

In the above discussion of quality control, as in all standard treatments of the subject, a CL basis is implied, the USL and LSL, for example, imply sharp cut-off boundaries within which the product is acceptable, otherwise it is not. However, natural events are not observed to have such exactitude, rather there are continuous gradations of attributes.

It will be recalled that FL sets generally do not exhibit such sharp boundaries, but may be formulated so that one set merges more gradually into adjacent sets. Standards and specifications almost invariably have an underlying CL structure and various devices are used to harmonise them with real data which does not have sharp categories. A certain amount of subjective judgement is also implied in their actual application.

Recasting standards and specifications in FL form removes the need for adjustments associated with sharp categories, enabling more flexible and realistic models to be applied. By the use of the logical conjunctive or disjunctive case the various factors in them may be merged to effect a single value judgement. This is beneficial where, for example, the combined effects of divers factors is important.

Consider two cases: the first is where the specification for a certain product parameter is expressed as bands. Let the be x, and assume for the moment that there are three bands in the range of x, as tabulated below.

Band	**Parameter**
A	$0<x<a$
B	$a<x<b$
C	$b<x$

It may be noted that the band changes discontinuously from A to B on x = a, and from B to C on x = b. On x = a and x = b the band is undefined.

In FL terms however, the bands would be expressed as illustrated in Figure 6.13.

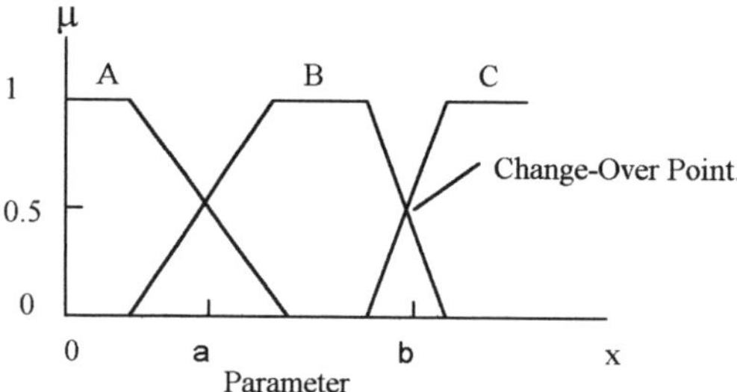

Figure 6.13 Typical Fuzzy Specification Banding

In the fuzzy specification banding illustrated in Figure 6.13, band A predominates up to parameter point x = a, beyond which point band B predominates up to point x = b, when Band becomes predominant. The FL description conveys more closely the meaning of the banding. A special case of this is where there is only one threshold.

A more complicated case is where there is more than one parameter which may act in tandem with another. It is then necessary to establish a rule -base for logical combination. Consider the case where there are two parameters, say y and z with a rule-base shown in Table 6.10:

Table 6.10 A Two-Parameter Rule-Base

	Y		
Z	A	B	C
R	Q_1	Q_2	Q_3
S	Q_4	Q_5	Q_6
T	Q_7	Q_8	Q_9

The Q_i entries in Table 6.10 control the joint effect of the y and z parameters in the specification. The rule-base is, in effect, a summary of the logical conjunctive case; IF Y AND Z THEN Q.

Example 6.11

In an agricultural country with a relatively low average disposable income there is a dairy sector with commercial milk production that is subject to regulation by a legal standard. The standard provides a reference for milk prices and producer advice for herd management, based upon compositional quality (CQ) and hygienic quality (HQ). The abbreviated specification from the standard is given below.

	Factor	**Standard**
i).	Butterfat (BF)	3% w/w (minimum).
ii).	Solids-not-fat (SNF)	8% w/w (minimum).
iii).	Total bacterial count (TBC)	<500 000 organisms/ml.
iv).	Somatic cell count (SCC)	< 1 million cells/ml.

For milk producer payments, the TBC and SCC are banded as follows:

	Band	**Thousands of organisms/ml.**
TBC	A	<50
	B	51-150
	C	151-250
	D	251-375
	E	376-500
	F	>500

	Band	**Thousands of organisms/ml.**
SCC	A	<300
	B	301-750
	C	751-1000
	D	>1000

Expert opinion has provided the following agreed rule-base:

Hygienic Quality: H-Rule-Base.

	TBC					
SCC	AT	BT	CT	DT	ET	FT
AS	E	E	G	M	P	P
BS	E	G	G	M	P	P
CS	G	M	M	P	P	P
DS	M	P	P	P	P	P

Compositional Quality: C-Rule-Base.

	SNF			
BF	AN	BN	CN	DN
AB	E	E	G	M
BB	E	G	M	P
CB	G	M	P	P
DB	M	P	P	P

Code: E = Excellent G = Good M = Moderate P = Poor

Evaluate a consignment of raw milk with the following laboratory test results:

3.0% BF 8.55% SNF 80 TBC (Thous./ml.) 765 SCC (Thous./ml)

Solution

Manipulation of the raw data is facilitated by normalising the data. For this purpose the following universe of discourse ranges are selected

SNF	8-13%	TBC	0-550 Thous.counts/ml.
BF	2.5-5.3%	SCC	0-1400 Thous.counts/ml.

The normalised diagrams for TBC,SCC,BF and SNF are shown in Figure Ex. 6.11.

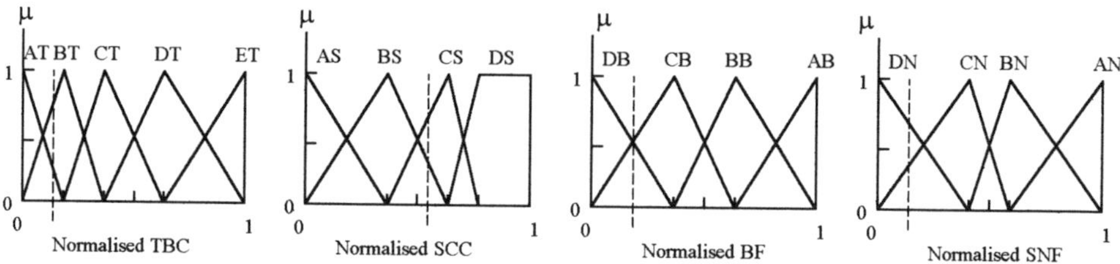

Figure Ex. 6.11 Partitioned Input Parameters

From the laboratory test data supplied, the normalised values are obtained as follows:

TBC = 0.145 SCC = 0.546 BF = 0.179 SNF = 0.110

These values are indicated by dashed vertical lines on the respective diagrams in Figure Ex. 6.11.

The TBC-SCC rule base and the membership values obtained from Figure Ex. 6.11 provide the following HQ conclusions:

IF TBC	AND SCC	THEN HQ	MIN	CONCLUSION
BT	CS	M	0.75,0.65	0.65 M
BT	BS	G	0.75,0.35	0.35 G
AT	CS	G	0.25,0.65	0.25 G
AT	BS	E	0.25,0.35	0.25 E

The overall conclusion is the union of the partial conclusions shown in the last column of the above tabulation. It is shown shaded in Figure Ex. 6.12 a).

The BF-SNF rule-base and the membership values which are also obtained from Figure Ex. 6.11 provide the following CQ conclusions:

IF BF	AND SNF	THEN CQ	MIN	CONCLUSION
DB	DN	P	0.5,0.75	0.5 P
DB	CN	P	0.5,0.25	0.25 P
CB	DN	P	0.5,0.75	0.5 P
CB	CN	P	0.5,0.25	0.25 P

Again, the overall conclusion is the union of the partial conclusions shown in the last column of the above tabulation and is shown in Figure Ex. 6.12 b).

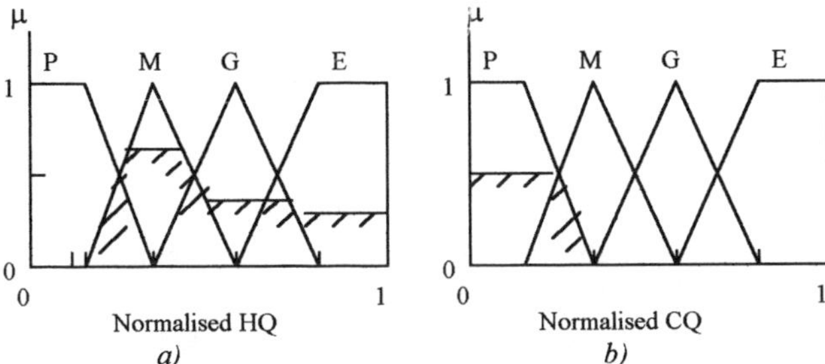

Figure Ex. 6.12 Conclusions for the Raw Milk Quality Evaluation of HQ and CQ

The conclusions in Figure Ex. 6.12 a) and b) may be defuzzified by the centroid method with the following results:

HQ diagram:	HQ value = 0.535	Membership values, 0.35 M, 0.65 G
CQ diagram:	CQ value = 0.175	Membership values, 0.20 M, 0.80 P

The conclusion is that the hygienic quality is mainly good, whilst the compositional quality is strongly poor.The evaluation process may be taken one stage further by merging the HQ and CQ results into an overall total quality (TQ) evaluation. This is accomplished by creating a rule-base using HQ and CQ as antecedents and TQ as the conclusion. A similar analysis to the above then yields the overall quality evaluation.

It may be noted in the above conclusions that the BF value occurs at the DB-CB change-over point, but this causes no problem, because a decision is not required as to which band to place this fact in. One other point is that the SCC was not considered to be as important as the TBC, and therefore the SCC was factored by 0.8 and the TBC by 1.2 (so that the total remained 2) to reflect the considered relative importance of 1.5:1. This is another benefit which can be built into the FL approach.

CHAPTER 7

MANAGEMENT SCIENCE

One of the major management functions is that of decision making based upon prior judgements (opinion forming), when a choice of various options is possible. The development of operational research (OR) (known as operations research in the USA) has been driven by the desire to reduce the subjective content of the decision making process and to replace it with quantitative rational methodology which provides a higher level of control for a range of various types of problems. This can lead to the maximum utility (better use of) limited resources.

Up to the present time, the methodology of OR broadly falls into two groups: mathematical programming and stochastic processes analysis, both resting on the foundation of CL. This has provided many successful examples of so-called scientific management and is now an established quantitative management tool. The advent of fuzzy logic offers the prospect of extending the methodology to include data that has an element of vagueness, which under the other treatments would be set aside. There are many texts available on the subject of OR dealing with it at various levels some are of a mathematical nature dealing with analytical models based on CL, but the origins of the subject were in the need to solve real problems of resource management. The pattern of application for this is:

i) Formulate the problem and the objectives
ii) Assemble data and identify the constraints
iii) Model the system
iv) Use the model, data and constraints to find solutions
v) Implement the solutions

Item v) is important and involves management problems of integration into the working system. This aspect is not considered further in this work. The purpose is to illustrate the solution of selected OR problems in terms of fuzzy logic and to indicate the way to more extensive applications of fuzzy OR methods.

7.1 Life-cycle costs

Periodic renewal of resources, particularly industrial and other equipment, is a universal requirement in all sectors of the economy and constitutes both a burden and an opportunity. Perceived renewal needs are motivated by:

i) Technological developments implying technical obsolescence
ii) Changes in functional requirements
iii) Completion of equipment useful working life

Item i) has become a very strong driving force in the industrial and commercial sectors of the more industrialised countries and is also important in other sectors such as health care and military. Item ii) can sometimes be offset by modification, or in the case of staff, by retraining. Needs due to items i) to iii) type considerations can be postponed by effective maintenance practices which can include upgrading and adapting equipment to delay technical obsolescence and to extend the useful working life of resources (which includes training of staff). It is clear therefore that maintenance, depreciation and replacement are closely inter-related themes, but they do have different relative values depending upon the geographical and temporal locations: monetary inflation and also currency exchange rates both influence replacement decisions, which can vary with the region and the economic cycle point.

It may be anticipated that comprehensive and precise analysis is not possible because of unmodelled factors in any analysis. Conclusions, especially about future trends, involve subjectivity and vagueness. This is where fuzzy logic becomes the natural mode of expression, as has been mention before in this text. High inflation manifests itself unpredictably in the medium to longer term, its effect on depreciation is that the residual market value of used equipment tends to be higher during periods of high inflation. Various inflationary models are possible.

If resources are depreciating in an inflationary economic climate, and assuming that the effects may be modelled as smooth differentiable functions, then for the present purposes a simple model of an equipment market value may be taken

$$C(T) = A \times T + B \times T^{-n} \qquad 7.1)$$

where fuzzy sets are implied for the coefficients A and B, also for the variables C and T. The range of T is $1 \leq T \leq T_1$ and T_1 is the useful lifetime of the resource. n is an index. The first term on the right-hand side of equation 7.1) reflects the influence of inflation on the level of prices, whilst the second term represents the resource depreciated value at a time T in a non-inflationary climate.

The minimum value of the resource is obtained when $dC(T)/dT = 0$, that is when

$$A - nB \times T^{-(n+1)} = 0 \qquad 7.2)$$

Therefore, $T_m = (nB/A)^{1/(n+1)}$ 7.3)

Equation 7.3) gives the time T_m for the minimum value to be achieved.

Using equation 7.3) to eliminate T in equation 7.1) gives the minimum value of C

$$C_m = (nA^n \times B)^{1/(n+1)}(1+(1/n^n)(B/A)^{(1-n)}) \qquad 7.4)$$

Example 7.1

An item of equipment is depreciating and its value in an inflationary climate at time T after initial commissioning is forecast to be given by the sum of two terms, C_1 and C_2, representing inflationary effects and annual depreciation

$$C_1(T) = BxT^{-0.2} \text{ Thous.GBP}$$

It is assumed that the additional value due to regional monetary inflation is represented to a first approximation by

$$C_2(T) = A \times T \text{ Thous.GBP}$$

where $A = [0//1.0+0.5//1.1+1.0//1.2+0.5//1.3+0//1.4]$

and $B = [0//30+0.5//35+1.0//40+0.5//45+0//50]$

Time T can be identified as a singleton [1.0//T] years.

Estimate the time taken to reach the minimum equipment value and also its value at this time.

Solution

From equation 7.3) the time set is given by

$$[0//t_1+0.5//t_2+1.0//t_3+0.5//t_4+0//t_5] = [[1.0//0.2]\times[0//1.0^{-1}+0.5//1.1^{-1}+1.0//1.2^{-1}+0.5//1.3^{-1}+0//1.4^{-1}] \times[0//30+0.5//35+1.0//40+0.5//45+0//50]]^{1/1.2}$$
$$= [0//6.00+0.5//6.3636+1.0//6.6667+0.5//6.9230+0//7.1429]^{0.8333}$$

where the Cartesian product of A and B are shown. The principal fuzzy time set is given by

$$T_m = [0//4.4508+0.5//4.6744+1.0//4.8592+0.5//5.0143+0//5.1468] \text{ years}$$

This set is illustrated in Figure Ex. 7.1

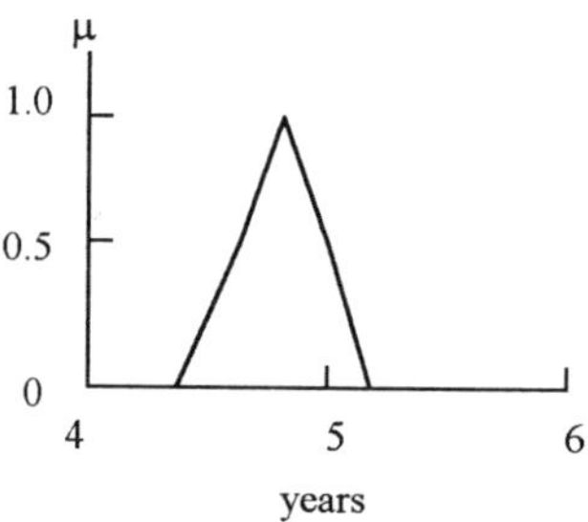

Figure Ex. 7.1 Fuzzy Time Set for Minimum Equipment Value

To find the minimum value, it is noted from equation 7.1) that

$$A\times T_m = [0//4.4508+0.5//5.1418+1.0//5.8310+0.5//6.5186+0//7.2055]$$

and $B\times T_m^{-0.2} = [0//21.618+0.5//25.354+1.0//29.156+0.5//33.057+0//37.090]$

Hence, $C_m = A\times T_m+B\times T_m^{-0.2}$

Retaining the principle set values of C_m yields

$$C_m = [0//26.069+0.5//30.496+1.0//34.987+0.5//39.576+0//44.300] \text{ Thous.GBP}$$

The minimum value set is shown in Figure Ex. 7.2

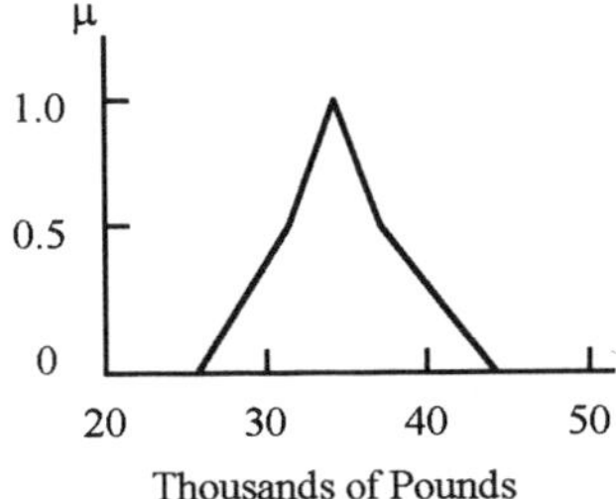

Figure Ex. 7.2 Fuzzy Minimum Value Set

It may be noted from Figures Ex. 7.1 and 7.2 that there is a significant spread of both the possible minimum value and the time at which this may occur. If n>1, which is usually the case, and if T_m >useful working life or the point of disposal, then obviously the resource will continue to diminish in value until disposed of. The solution in terms of fuzzy sets does give an idea of the uncertainty in this. Both diagrams may be defuzzified to yield the strongest case, which is 35 thousand GB pounds in 4.85 years.

Various models of inflation are possible in addition to the one used above, for example, a common one is based on the reduction of interest rate by the inflation rate to achieve an inflation corrected rate.

The notional value of a resource at some point in the future depends upon a number of factors, amongst the more important is the quality and volume of maintenance invested in the resource during its working life up to that point from the time of installation. This is determined by the effectiveness of maintenance management. A rough guide is the maintenance investment, which is provided by regular payments into a sinking fund.

Suppose that maintenance payments are made at a flat rate of M, the integrated effect of these payments at a later time T , based upon a uniform interest rate of I would be

$$F = \int_0^T M \exp I \times T' dT' \qquad 7.5)$$

$$= M \times I^{-1} \times [\exp I \times T - 1]$$

where T' is non-current time.

This is similar in effect to the expression

$$F = [M \times I^{-1}] \times [(1+I)^n - 1]$$

which is the sum of a geometric series which is found in some texts. It is more accurate for small values of time, but rapidly converges to the exponential expression.

The present value F_p of the investment given in equation 7.5), as will be shown a posteriori, is given by

$$F_p = F \times \exp\text{-}I \times T \qquad 7.6)$$

$$= M \times I^{-1} \times [1\text{-}\exp\text{-}I \times T]$$

This is similar in effect to the expression

$$F = M \times I^{-1} \times [1\text{-}(1+I^{-n})]$$

which the sum of another geometric series. This is also more accurate for small values of time and converges to the exponential expression.

Future values are not strictly comparable if made at different times. They are therefore frequently rebased to current values by inverting the effect of compound interest as follows.

Let there be a continuous flow of interest which has a rate that is proportional to the current value of the investment

Then, $$dV/dT = I \times V \qquad 7.7)$$

Integrating this expression and applying the condition that the initial value (the principal) is P at time T = [1.0//0], yields

$$V(T) = P \times \exp I \times T \qquad 7.8)$$

where I is the fuzzy interest set. Equation 7.8) is similar in effect to the compound interest formula

$$V(n) = P(1+i)^n$$

The current value of a value V(T) at some time in the future is therefore

$$P = V(T)\exp\text{-}I \times T \qquad 7.9)$$

This result may be applied to the depreciation expression, equation 7.7). The apparent present value of a future value may then be expressed as

$$U(T) = C(Y) \times \exp\text{-}I \times T \qquad 7.10)$$

or, $$U(T) = [A \times T + B \times T^{-n}] \times \exp\text{-}I \times T \qquad 7.11)$$

By differentiating equation 7.11) with respect to T it may be shown that the minimum value of C(T) may again be found from equation 7.3), if it exists within the selected interval.

Example 7.2

A hospital requires a supply of hot water and both gas-fired and electrode boilers are available that have the service capacity. The gas-fired boiler has a capital cost of 6 800 GBP and an annual fuel and maintenance cost of 800±15% GBP. The electrode boiler has a capital cost of 1 300 GBP and an annual cost of 1 500±12%. It is estimated that the current interest rate of 9% may rise within the limits of 11% and 6% in the nest five year period It is assumed that the effect of inflation on the depreciated plant value may be given by an additive term (AxT) which is proportional to the plant age, the proportionality coefficient being 20.3±15% GBP per annum. The depreciated equipment value is given by $BT^{-0.2}$

Find the nett present value of the life-cycle cost at the end of each year during the first five-year operating period. Neglect plant installation costs.

Solution

In the following solution only the principle elements of the Cartesian products will be considered, assuming that the depreciated value of the equipment is given by

$$C = A\times T + B\times T^{-0.2}$$

where, A = [0//20+0.5//22+1.0//23+0.5//24+0//26]

For the electric boiler, B_E = [0//1100+0.5//1200+1.0//1300+0.5//1350+0//1400]
For the gas boiler, B_G = [0//6400+ 0.5//6600+1.0//6800+0.5//7000+0//7200]

The interest fuzzy set is I = [0//6+0.5//7.5+1.0//9+0.5//10+0//11] %

The time at five years is expressed as a singleton; [1.0//5]. Considering the expression for C, the following results are obtained:

$A\times T$ = [0//100+0.5//110+1.0//115+0.5//120+0//130]
$B_E\times T^{-0.2}$ = [0//797+0.5//870+1.0//942+0.5//978+0//1014]
$B_G\times T^{-0.2}$ = [0//4637+0.5//4784+1.0//4929+0.5//5074+0//5219]
$I\times T$ = [0//30+0.5//37.5+1.0//45+0.5//50+0//55]
exp-$I\times T$ = [0//0.7408+0.5//0.6873+1.0//0.6376+0.5//0.6065+0//0.5769]

The present value of the depreciated future value of the equipment is given by equation 7.11), from which the following sets are obtained:

U_E = [0//664+0.5//674+1.0//674+0.5//666+0//661] GBP
U_G = [0//3511+0.5//3364+1.0//3210+0.5//3150+0//3080] GBP

where U_E and U_G are the electric and gas boiler values respectively. The present values (H_E,H_G) of the annual cumulative costs to time T are

$$H_E = F_E \times I^{-1} \times [1\text{-exp-}I \times T]$$
$$= [0//5702+0.5//5879+1.0//6040+0.5//6257+0//6462] \text{ GBP}$$

and

$$H_G = F_G \times I \times [1\text{-exp-}I \times T]$$
$$= [0//2938+0.5//3085+1.0//3221+0.5//3384+0//3539] \text{ GBP}$$

Cumulative present values of the total disbursement for 5 years are found by adding the initial equipment capital costs of 1 300 and 6 800 for the two plants to the H_E and H_G sets respectively, giving the following results:

$$L_E = [0//7002+0.5//7179+1.0//7340+0.5//7557+0//7762] \text{ GBP}$$

and

$$L_G = [0//9738+0.5//9885+1.0//10021+0.5//10184+0//10139] \text{ GBP}$$

The present value life-cycle costs , (J_E and J_G) to five years are found by subtracting the present value of the equipment future market values, (U_E and U_G), from L_E and L_G respectively, giving

$$J_E = [0//6338+0.5//6505+1.0//6666+0.5//7034+0//7053] \text{ GBP}$$

and

$$J_G = [0//6227+0.5//6521+1.0//6811+0.5//7034+0//7053] \text{ GBP}$$

These two fuzzy sets are illustrated in Figure Ex. 7.3. The maximum membership values for the two types of plant indicates that the gas boiler is slightly more expensive than the electrical boiler over a five year period. Figure Ex. 7.3 also indicates that there is a similarity between the two costs. The similarity may be quantified by considering the ratio of the intersection of J_E and J_G with their union. By simple calculation this is about 0.74, and the dissimilarity is 1-0.74, or 0.26. This sufficient to favour the electrode boiler.

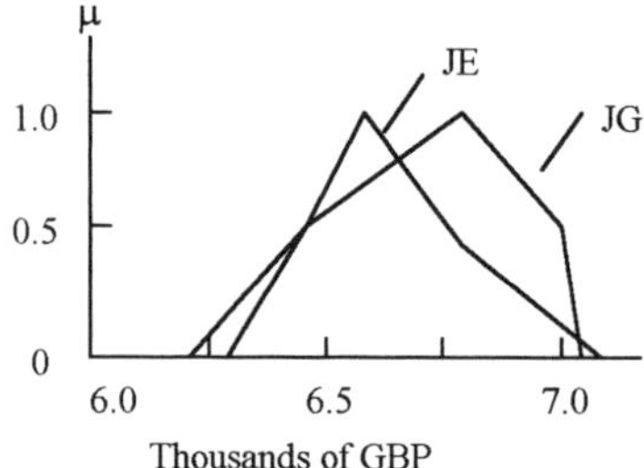

Figure Ex. 7.3 Present Value Life-Cycle Costs to Year Five

7.2 Plant amortisation

The previous section compared two types of boiler to satisfy a particular service requirement. Another related consideration of asset management is that of deciding the most favourable point in time to replace resources, such as fixed installations. If financial considerations only are of importance, the most favourable point is often judged by the minimum of the repeated payments figure associated with a given effective interest rate.

Consider a case of future payments made at a uniform rate, P. The cumulative value Q(T) at time T of payments made at earlier times (T') with an effective interest rate, I, is given by

$$Q(T) = P\int_0^T \exp\text{-}I\times T'dT' \qquad 7.12)$$
$$= P\times I^{-1}[1\text{-}\exp\text{-}I\times T]$$

Conversely, the uniform payments rate, P, that would provide the term sum Q(T) is from this expression to be

$$P(T) = Q(T)\times I\times[1\text{-}\exp\text{-}I\times T]^{-1} \qquad 7.13)$$

Although P(T') is constant, P(T) is a variable function of T; the uniform rate depends upon T.

The replacement criterion may be stated as, dP(T)/dT = 0.

Generally, the replacement point, determined by the above criterion, is more conveniently evaluated numerically.

Example 7.3

The estimated projected seven-year plant resale present value is given in Table Ex. 7.1 and the present value of the capital,maintenance and other annual costs for the plant is given in Table Ex. 7.2. Prepare an analysis of the most favourable financial period for plant replacement, based upon the above replacement criterion.

Table Ex. 7.1 Plant Resale Present Value

	Membership Value				
Year	0	0.5	1.0	0.5	0
1	8670	9151	9633	10115	10596
2	5833	6157	6481	6805	7179
3	3893	4109	4325	4541	4758
4	3124	3297	3471	3645	3818
5	2459	2593	2729	2865	3002
6	1878	1983	2087	2191	2296
7	1379	1455	1532	1609	1685

Table Ex. 7.2 Cumulative Value of Investment

	Membership Value				
Year	0	0.5	1.0	0.5	0
1	12947	13666	14385	15248	16111
2	13370	14113	14856	15747	16389
3	13857	14627	15397	16321	17245
4	14749	15569	16388	17371	18354
5	17082	17082	17981	19060	20139
6	18061	19064	20068	21272	22476
7	20474	21611	27748	24112	25478

Assume an effective flat interest set, I = [0//0.06+0.5//0.075+1.0//0.09+0.5//0.10+0//0.11]

Solution

Plant resale values may be regarded as credit and cash spent as debit, the balance is the nett present value of the life-cycle costs up to a given point in time. In the following work, only principal set values are shown. The nett present value of the life-cycle costs are calculated using equation 7.12 and the above data. The results are shown below in Table Ex. 7.3.

Table Ex. 7.3 Nett Present Value of Life-Cycle Costs. (Q(T))

	Membership Value				
Year	0	0.5	1.0	0.5	0
1	4277	4515	4752	5133	5515
2	7537	7956	8375	8942	9260
3	10164	10518	11072	11780	12487
4	11625	12272	12917	13726	14536
5	14623	14484	15252	16195	17137
6	16183	17081	17981	19081	20180
7	19095	20156	21216	22503	23793

The values in table Ex. 7.3. may be used in conjunction with equations 7.12 and 7.13. The exponential term $[1\text{-exp-}I\times T]\times I^{-1}$ is set out below in Table 7.4.

Table Ex. 7.4 Values of $[1\text{-exp-}I\times T]\times I^{-1}$

	Interest Set (I)				
T years	0.06	0.075	0.09	0.10	0.11
1	0.9700	0.9640	0.9567	0.9520	0.9473
2	1.8550	1.8574	1.8300	1.8130	1.7954
3	2.7450	2.6667	2.6289	2.5920	2.5555
4	3.5567	3.4560	3.3589	3.2910	3.2364
5	4.3200	4.1693	4.0267	3.9350	3.8464
6	5.0387	4.8316	4.6361	4.5119	4.3922
7	5.7159	5.4459	5.1934	5.0341	4.8817

The payments rate (P(T)) in equation 7.13 may therefore be found directly from the above data, the results are shown below:

Table Ex. 7.5 Tabulation of Payment Rates P(T)

	Membership Value				
Year	0	0.5	1.0	0.5	0
1	4409	4684	4967	5392	5820
2	4063	4284	4577	4923	5157
3	3703	3915	4212	4545	4886
4	3268	3551	3846	4171	4491
5	3385	3474	3788	4116	4455
6	3790	3535	3878	4229	4595
7	3914	3701	4085	4470	4874

It may be noted that the P(T) values in the above table appear to reach a minimum around year 4. It would therefore be advantageous to replace the plant at about this point.

7.3 Tender adjudication

Purchasing goods or services by tender has become more common in recent years, but in some sectors such as civil engineering and construction, government and armed forces administration it has a long established history. The procedure for large projects is very elaborate, whilst at the other end of the scale, for minor civilian contracts the conditions and criteria may be very simple with little documentation and adjudicated purely on price.

For tenders which are adjudicated on other criteria in addition to price, an assessment system is required that will compare submitted tenders on a number of selected factors and which will present the adjudication team with a single quantitative metric for each tender and which will provide a reliable and robust basis on which plausible decisions may be made.

The treatment given below is a structured judgement system based upon ranking scores of the tender criteria and finding the intersection of this set with the chosen set of priorities. In formal terms this is expressed as

$$s = \Sigma_{c,p}\mu_c \wedge \mu_p \quad 7.14)$$

where μ_c and μ_p are the membership values of the tender criteria and the priorities respectively.

A normalised adjudication metric may be found by dividing the value, s, by the maximum possible intersection, m, thus

$$n = s/m \quad 7.15)$$

The maximum possible intersection is that of μ_c with itself

$$m = \Sigma_c \mu_c \wedge \mu_c \quad 7.16)$$

Example 7.4

A civil engineering aid project has been let out to tender and it is estimated from prior experience that a reasonable cost for the project would be about 10 million GBP to produce an equivalent construction of the desired quality. For this project the value criteria and their priorities have been selected by the customer management team and consultants to be as follows:

	Criterion	**Relative Priority Rating**
1.	Tender sum (TS)	0.9
2.	Proposer's relevant experience (RE)	1.0
3.	CVs of senior supervisory personnel (CV)	0.9
4.	Proposed programme of works (PW)	0.9
5.	Tender variations/Qualifications/ Alternatives offered (TV)	0.8
6.	Basic price list. (Check on Proposer's costings) (PL)	0.7

The above ratings are on a scale; 0-1.0
Three properly presented tender proposals have been received; A,B and C. The customer management team has considered these three tenders and agreed the following score table on a scale; 0-1.0.

Tender Proposal	**Criterion Score**					
	TS	RE	CV	PW	TV	PL
A	0.78	0.3	0.3	0.1	0.4	0.1
B	0.87	0.9	0.8	0.8	1.0	0.9
C	0.96	0.7	0.8	0.9	1.0	0.9

Evaluate the adjudication metrics.

Solution

The relative priority ratings of the selected criteria are shown in Figure Ex. 7.4.a), b) and c). The scores for the three submitted tenders A,B and C are superimposed on the diagrams for comparison. The comparison is measured by the intersection of each pair of sets. Consider Figure Ex. 7.4.a), the intersection is given by equation 7.14)

$$s = \Sigma \min(\mu_A, \mu_P)$$
$$= 0.78+0.3+0.3+0.1+0.4+0.1 = 1.98$$

The maximum possible is the self-intersection, equation 7.16)

$$m = 3*0.9+1.0+0.8+0.7 = 5.3$$

Hence, from equation 7.15) the correlation metric for tender A is

$$n = 1.98/5.3 = 0.381$$

In the same way, the metrics for tenders B and C may be found. the results are as follows

Tender	**Correlation Metric**
A	0.381
B	0.937
C	0.935

On the above basis, the metrics B and C are virtually indistinguishable. To discriminate between them the customer management team would need to select supplementary criteria for a tie-breaker adjudication.

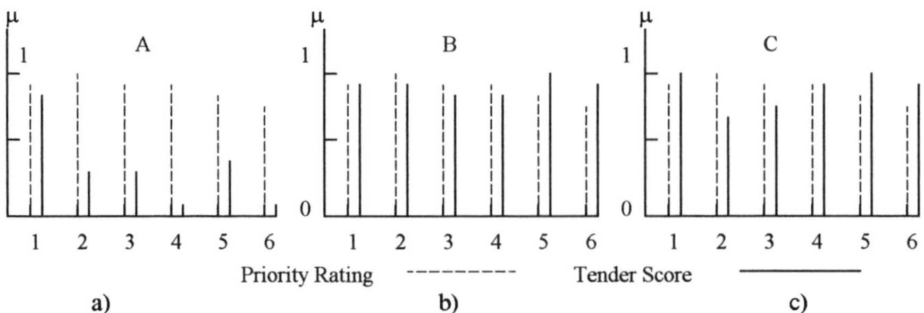

Figure Ex. 7.4 Comparison of Priority Ratings with Tender Scores

7.4 Staff selection process

The context of the following discussion is that of a human resources function of a corporate community. In this context, the function is exercised within a framework of corporate knowledge and experience, and the process will be based upon knowledge engineering in which the principles of fuzzy logic are applied. Information may be gathered in various ways, but judgements have to be made and decisions reached on information which often is not amenable to deterministic analysis. The following discussion shows how FL methods aid the process.

The selection process considered here comprises five operations, viz.:

i) Choice and publication of vacancy criteria; Prioritisation and quality scale; Relational base
ii) Coarse filtering of applicants to provide a qualified candidates list
iii) Candidate evaluation by relational base composition
iv) Interviews and listing of pair comparisons
v) Array of pair comparisons and final ranking

The choice of vacant post criteria, also prioritisation and the quality scale are a management and expert team function involving a degree of subjectivity. Conclusions are reached by consensus. Coarse filtering by an expert team provides a qualified candidates list. Candidate evaluation is a mechanical process which is described below. Pair comparisons are management and expert team value judgements concluded by consensus opinion. The comparison array and final ranking are mechanical processes which are also described below.

An initial screening process will filter out all those candidates failing to satisfy the publicised requirements in one or more ways. The requirements would typically include:

i) Minimum educational achievement (ED)
ii) Relevance of qualifications (RE)
iii) Period of relevant experience (EX)
iv) Status of previous post (ST)
v) Benefits required by the candidate (BE)

Such a list would be amended, depending upon the exigencies of the post and the management.The management team then prioritises these criteria by defining membership values to provide a fuzzy set

$$P = [\mu_{ED}/ED+\mu_{RE}/RE+\mu_{EX}/EX+\mu_{ST}/ST+\mu_{BE}/BE] \qquad 7.17)$$

The quality scale may be defined in terms of a fuzzy set

$$Q = [\mu_{PR}/PR+\mu_{MD}/MD+\mu_{GD}/GD+\mu_{SP}/SP] \qquad 7.18)$$

where PR = Poor; MD = Moderate; GD = Good; SP = Superior.

If expressions 7.17) and 7.18) above define agreed standards, then a relational array may be found by their Cartesian product. This array may be applied to evaluate all those candidates deemed to be qualified. The relational array is expressed as

$$R = Q \times P = \begin{array}{c|cccc} & PR & MD & GD & SP \\ ED & \mu_{11} & \mu_{12} & \mu_{13} & \mu_{14} \\ RE & \mu_{21} & \mu_{22} & \mu_{23} & \mu_{24} \\ EX & . & . & . & . \\ ST & . & . & . & . \\ BE & . & . & . & \mu_{54} \end{array} \quad 7.19)$$

The evaluation of each candidate is obtained by consensus scoring of each of them by the selection team, giving a fuzzy set, W

$$W = [\mu_{ED}//ED+\mu_{RE}//RE+\mu_{EX}//EX+\mu_{ST}//ST+\mu_{BE}//BE] \quad 7.20)$$

The evaluation (E = WoR) is then given by

$$WoR = \text{maxmin } |\mu_{ED}\ \mu_{RE}\ \mu_{EX}\ \mu_{ST}\ \mu_{BE}| \begin{vmatrix} \mu_{11} & \mu_{12} & \mu_{13} & \mu_{14} \\ \mu_{21} & \mu_{22} & \mu_{23} & \mu_{24} \\ . & . & . & . \\ . & . & . & . \\ . & . & . & \mu_{54} \end{vmatrix} \quad 7.21)$$

$$= |\mu'_{PR}\ \mu'_{MD}\ \mu'_{GD}\ \mu'_{SP}|$$

The fuzzy set evaluation for a particular candidate is therefore given by

$$E = [\mu'_{PR}//PR+\mu'_{MD}//MD+\mu'_{GD}//GD+\mu'_{SP}//SP] \quad 7.22)$$

The next process is the comparison of candidate pairs by the management and expert team by interview. The results of the above evaluation together with the interview are the basis for the comparison on a scale of 0 to 1. The comparisons are arranged as an array

$$C = \begin{array}{c|cc} & A & B \ldots\ldots\ldots\ldots N \\ A & 1 & ab \ldots\ldots\ldots\ldots an \\ B & ba & 1 \ldots\ldots\ldots\ldots bn \\ . & . & \ldots\ldots\ldots\ldots\ldots \\ . & . & \ldots\ldots\ldots\ldots\ldots \\ N & na & nb \ldots\ldots\ldots\ldots 1 \end{array} \quad 7.23)$$

In the above array, if the comparisons were non-subjective then ba = 1-ab, but in general, to accommodate a degree of subjectivity, this condition is not applied. Ranking of the comparisons is obtained by finding the array of the ratios of complementary pairs, defined by

$$\mathit{ij} = ij/ji \quad 7.24)$$

The comparison array (*C*) of the *ij* ratios achieved by this enables the best (highest) value in each row to be found by inspection. These best values may then be ranked by inspection in descending order. Equivalently, the worst (lowest) value in each column may be found by inspection and these may be ranked in descending order. The best of the best is the same as the worst of the worst.

The ordering of the candidates is therefore achieved and a final choice may be made.

Example 7.5

A short-list of candidates is drawn up from the applicants for an advertised post in a large corporation. The management and expert team responsible have specified the following criteria for the post:
i) Minimum educational achievements (ED)
ii) Professional experience (RE)
iii) Status of previous post (PP)
iv) Referees comments (RC)

The team has prioritised the above criteria, giving the following set

$$P = [0.7//ED+0.8//RE+0.7//PP+0.6//RC]$$

A quality template has also been established as follows

$$Q = [0.4//PR+0.6//MD+0.8//GD+1.0//SP]$$

This means: 0 to 0.4 is Poor, 0.4 to 0.6 is Moderate, 0.6 to 0.8 is Good and 0.8 to 1.0 is Superior.

a) The score of each of four candidates, A to D in each of the four prioritising criteria is

	ED	RE	PP	RC
A =	0.6	0.7	0.9	0.4
B =	0.5	0.9	0.6	0.8
C =	0.6	0.6	0.8	0.6
D =	0.4	0.7	0.5	0.6

Evaluate each of the four candidates.

b) Following an interview and taking into consideration the evaluations found previously, a pair comparisons (C) array was established

		A	B	C	D
	A	1	0.38	0.61	0.35
C =	B	0.72	1	0.36	0.56
	C	0.45	0.85	1	0.38
	D	0.90	0.96	0.18	1

Find the outcome of the ranking process for the candidates.

Solution

a) The relational array is given by equation 7.19)

R = Q×P

	PR	MD	GD	SP
ED	0.4	0.6	0.7	0.7
RE	0.4	0.6	0.8	0.8
EX	0.4	0.6	0.7	0.7
ST	0.4	0.6	0.6	0.6

The evaluation array is given by equation 7.21)

E = W∘R

	PR	MD	GD	SP
E_A =	0.4	0.6	0.7	0.7
E_B =	0.4	0.6	0.8	0.8
E_C =	0.4	0.6	0.7	0.7
E_D =	0.4	0.6	0.7	0.7

From the above array, the best evaluation is for candidate B, the other three are equal.

b) From the given data in the pair comparisons array the ratios of the complementary pairs may be found using equation 7.24). The resulting (C) array is shown below

	A	B	C	D
A	1	0.53	1.36	0.38
B	1.89	1	0.42	0.58
C	0.74	2.36	1	2.11
D	2.57	1.71	0.47	1

Ranking the rows according to the highest value in each gives: (2.57,2.11,1.89,1.36). In descending order the rows are therefore ranked: (D,C,B,A). Alternatively, ranking according to the lowest value in each column gives: (0.38,0.42,0.53,0.74). The ranking is as before.

7.5 Supplier selection

The management team of an enterprise often has to choose between alternative sources for the satisfaction of needs for goods or services. The choice is often made on the basis of perceived levels of quality of product, quality of service and price. Each of these is to be evaluated and the outcomes merged into a final judgement, all of which are within the scope of FL methods. Quality of service and quality of product are much more subjective than price, which usually appears to be clear-cut. Product and service quality considerations are treated in Section 7.6.

The conventional analysis and presentation of information of the pricing aspect based upon CL and ignoring vagueness and ambiguity in the known data will obscure some of the possibilities in solutions to problems, and these may influence judgements and therefore decision making. However, FL through the application of the extension principle and other means enables some of the conventional methods to be adapted to the more articulate FL treatment, and solutions are revealed with greater depth than otherwise.

Example 7.6

A manufacturing company is planning the production of a new product of limited production volume, for which a certain component is required. The planning team has identified three possible options which may satisfy the delivery and specification requirements, these are as follows:

i) Purchase a new machine that would result in the production of the components at a cost of about 0.80 GBP±12.5% per unit, plus overhead costs of 300 GBP per week.
ii) Purchase an alternative new machine that would result in the production of the components at a cost of about 0.40 GBP ±10% per unit, plus overhead costs of 420 GBP per week.
iii) Sub-contract (outsource) the demand for a delivered price of 2.10 GBP per unit for volumes of 200-300 per week.

Due to uncontrolled factors, the in-house production rate may, on average, fluctuate by about ±10%.. The planning team requires the following information:

a) The break-even point between production with machine option i) and the sub-contracting option iii).
b) The possibility that machine options i) and ii) will produce the same volume of components at the same cost.

The planning team has established the following information.

Components per week. X = [0//0.9x+0.5//0.95x+1.0//x+0.5//1.05x+0//1.10x]
Unit costs: Machine i) A = [0//0.70+0.5//0.75+1.0//0.80+0.5//0.85+0//0.90]
Machine ii) B = [0//0.36+0.5//0.38+1.0//0.40+0.5//0.42+0//0.44]

Solution

a) Expressing the relationships in terms of fuzzy sets. Let C = [1.0//300] be the overhead cost of the first new machine and let W = [1.0//2.0] be the sub-contract unit cost. It is clear that the break-even point is governed by the condition

$$C+[A\times X] = W\times X$$

The Cartesian product, A×X is given in the following tabulation

A \ **X**	0//0.9	0.5//0.95	1.0//1.0	0.5//1.05	0//1.1
0//0.70	**0.630**	0.665	0.700	0.735	0.770
0.5//0.75	0.675	**0.713**	0.750	0.788	0.825
1.0//0.80	0.720	0.760	**0.800**	0.840	0.880
0.5//0.85	0.765	0.808	0.850	**0.893**	0.935
0//0.90	0.810	0.835	0.900	0.945	**0.990**

The principal set in the above array is shown in bold type. There is a corresponding array for the associated minimum membership values which is not shown here. The Cartesian product may therefore be expressed as

A×X = [0//0.630+0.5//0.713+1.0//0.800+0.5//0.893+0//0.990]×X

Also, W×X = [1.0//2.0]x[0//0.9+0.5//0.95+1.0//1.0+0.5//1.05+0//1.10]×X

= [0//1.80+0.5//1.90+1.0//2.00+0.5//2.10+0//2.20]×X

Now, C = [W-A]×X =

[0//(2.20-0.63)+0.5//(2.10-0.713)+1.0//(2.00-0.80)+0.5//(1.90-0.893)+0//(1.80-0.99)]×

X

Hence, X = C×[W-A]$^{-1}$ = [1.0//300]x[0//1.57+0.5//1.39+1.0//1.20+0.5//1.01+0//0.87]
= [0//219+0.5//233+1.0//250//+0.5//271+0//292]

Note that by turning set A around from end-to-end an alternative set could be obtained which is contained in the given set.

The fuzzy set X above expresses the break-even point for case a). It is illustrated in Figure Ex.7.5.

b) In this case the break-even point between option i) and option ii) is required. This is governed by the condition

C+A×X = D+B×X

where B = [1.0//420], and X is the new break-even point.

The Cartesian product, B×X is given in the following tabulation

	X				
B	0//0.9	0.5//0.95	1.0//1.0	0.5//1.05	0//1.1
0//0.36	**0.324**	0.342	0.360	0.378	0.396
0.5//0.38	0.342	**0.361**	0.380	0.399	0.418
1.0//0.40	0.360	0.380	**0.400**	0.420	0.440
0.5//0.42	0.378	0.399	0.420	**0.441**	0.462
0//0.44	0.396	0.418	0.440	0.462	**0.484**

The principal set in the above array is shown in bold type. There is a corresponding array for the associated membership values which is not shown here. The Cartesian product may therefore be expressed as

B×X = [0//0.324+0.5//0.361+1.0//0.400+0.5//0.441+0//0.484]×X

Now, D-C = [A-B]×X =
[0//(0.63-0.484)+0.5//(0.713-0.441)+1.0//(0.8-0.4)+0.5//(0.893-0.361)+0//(0.99-0.324)]×X
= [0//0.146+0.5//0.272+1.0//0.400+0.5//0.532+0//0.666]×X

Hence, X = [D-C]×[A-B]$^{-1}$ = [1//(420-300)]x[0//6.849+0.5//3.676+1.0//2.500+0.5//1.880+0//1.502]
= [0//180+0.5//226+1.0//300+0.5//441+0//822]

Note that by turning B around end-to-end another set could be obtained which is contained in the given set.The fuzzy set X above expresses the break-even point for case b). It is also illustrated in Figure Ex. 7.5.

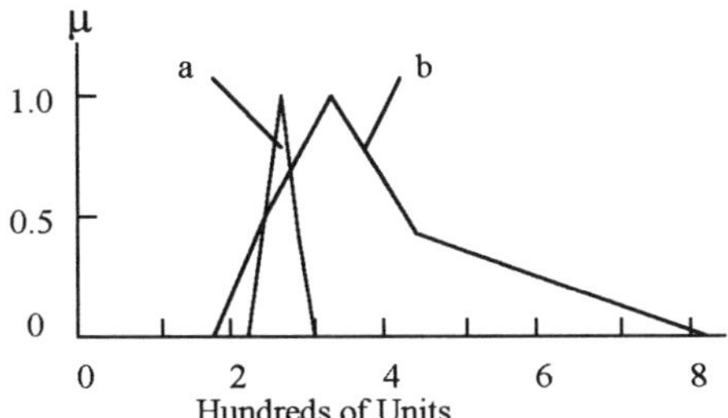

Figure Ex. 7.5 Weekly Production Break-Even Point
a) Comparing Options i) and ii)
b) Comparing Options i) and iii)

It may be noted that by inspection of the intersection of cases a) and b) shown in Figure Ex. 7.5, there is some possibility of producing the same number of components for the same cost using options i) and ii) when the production rate is between about 220 to 300 units per week. This would not have been revealed by a deterministic analysis. Such an analysis would have indicated 250 components for case a) and 300 components for case b).

7.6 Product and service quality

It is becoming increasingly accepted that product quality and service quality are only definable by the consumer. On the local, national and global levels there is such a great variety of products and services and also of consumers that general methods of subjective value judgements would arguably be impossible to find, and segmentation on both the demand and supply sides is needed to reduce the problems to manageable proportions. The consumer may be internal or external to a given organisation. In the present climate of continuous improvement and change, there is an increasing need to develop tools by which the changes can be monitored and judged.

The objective here is to outline a plausible and robust method for the evaluation of an economic object (artefact or service). The method espoused is that of creating a consumer standard value judgement bank for the genre (VJB). The VJB is accessed by composition with a given set of specimen critical attributes.

As an example of the critical attributes of an artefact, the following list might be used:
i) Critical dimensions conformance (CD)
ii) Materials specification (MS)
iii) Surface finish (SF)
iv) Inspection schedule (IS)
v) Product test programme (TP)

The corresponding value judgements might be
i) Excellent (E). ii) Very good (V). iii) Good (G). iv) Fair (F). v) Poor (P).

Similarly, for service the critical attributes might be
i) Reliability and delivery time (R)
ii) Workforce skills (W)
iii) Organisation efficiency (O)
iv) Management cooperativeness (M)
v) Product improvement proactivity (P)

The corresponding value judgements might be
i) Excellent (E). ii) Superior (S). iii) Adequate (A). iv) Inferior (I).

The universe of discourse of quality of that particular genre is the collection of critical attributes. Each attribute is scored in the range, 0 to 1.0 by each member of the expert consumer panel and added to the VJB array, so that the array entry is the sum of the panel scores for that specific attribute. The array is then normalised so that the sum of the entries in each row is unity. The generality of the VJB is increased by polling the expert consumer panel on a number of other specimens of the same genre. Let the arrays so found be, $R_1, R_2 ..... R_n$, then the resultant VJB obtained by combining these samples is

$$R = R_1 \cup R_2 .... \cup R_n \qquad 7.25)$$

For example, suppose that there are two specimens of a genre with three attributes and three value judgements. Let the panel conclusions be as shown below in the VJB Tables 7.1 and 7.2.

Table 7.1 Specimen 1 VJB

	E	S	A
A	0.26	0.58	0.16
B	0.11	0.40	0.49
C	0.00	0.66	0.34

Table 7.2 Specimen 2 VJB

	E	S	A
A	0.54	0.00	0.46
B	0.35	0.30	0.35
C	0.36	0.33	0.31

The intersection of these two arrays is shown in Table 7.3, and the corresponding normalised VJB values are shown in Table 7.4.

Table 7.3 Combined VJB

	E	S	A
A	0.26	0.00	0.16
B	0.11	0.30	0.35
C	0.00	0.33	0.31

Table 7.4 Normalised Combined VJB

	E	S	A
A	0.62	0.00	0.38
B	0.15	0.39	0.46
C	0.00	0.52	0.48

The likely effect of a change in a good (or service) is found by evaluating the attributes of the good which appear in the VJB, each on a scale of 0 to 1.0, and to produce a value judgement by composition with the VJB.

Example 7.7

A panel of ten expert consumers has provided value judgements about two different products of a genre. There are five critical quality attributes (Q) and five value judgement categories (J). The total score of the panel for each attribute is shown below in Table Ex. 7 6 and Table Ex. 7.7:
Table Ex. 7.6 VJB Score for Product 1

		J				
	Q	E	V	G	F	P
	A	4.6	5.6	5.5	1.8	3.1
R =	B	8.4	3.4	1.0	4.5	7.9
	C	1.8	8.6	6.5	9.7	2.4
	D	6.8	5.4	10.0	4.1	3.5
	E	3.9	6.5	8.9	0	8.7

Table Ex. 7.7 VJB Score for Product 2

		J				
	Q	E	V	G	F	P
	A	4.7	5.8	6.4	8.9	7.1
S =	B	7.0	4.5	9.4	2.7	8.8
	C	1.0	0	1.9	0	7.1
	D	7.7	7.9	1.8	2.6	1.0
	E	0	6.9	6.2	0	5.0

A prototype of a third product of the same genre becomes available at a later date and it is has been estimated to have the following quality profile

X = [0.47//A+0.09//B+0.38//C+0.22//D+0.95//E]

The existing product has a J-grade of G. Compare the prototype grade with this.

Solution

The normalised VJB arrays for product 1 and product 2 are, respectively

		J				
	Q	E	V	G	F	P
	A	0.22	0.27	0.26	0.10	0.15
R' =	B	0.33	0.15	0	0.20	0.32
	C	0.10	0.28	0.20	0.32	0.10
	D	0.22	0.18	0.33	0.14	0.13
	E	0.14	0.23	0.32	0	0.31

		J				
	Q	E	V	G	F	P
	A	0.14	0.18	0.19	0.27	0.22
S' =	B	0.22	0.14	0.27	0.10	0.27
	C	0.10	0	0.19	0	0.71
	D	0.37	0.38	0.11	0.14	0
	E	0	0.38	0.34	0	0.28

The resultant VJB (T') is obtain by the intersection of R' and S', normalising by rows, thus

		J				
	Q	E	V	G	F	P
	A	0.18	0.24	0.25	0.13	0.20
T' =	B	0.30	0.19	0	0.14	0.37
	C	0.26	0	0.49	0	0.25
	D	0.34	0.28	0.17	0.21	0
	E	0	0.28	0.39	0	0.32

The resultant value judgement (W) is given by the composition of P with T'

$$W = X \circ T'$$

By inspection, this is given as

W = [0.26//E+0.28//V+0.39//G+0.21//F+0.33//P]

This fuzzy set represents the J-grade of the prototype, to be compared with the grade of G for the existing product. The comparison is facilitated by finding the centroid of the fuzzy set. No weights are given for the valuation categories, so it is assumed that they are all of equal weight. The centroid (Y) is found by taking moments and dividing by the sum of the membership values

$$Y = (1*0.26+2*0.28+3*0.39+4*0.21+5*0.33)/(0.26+0.28+0.39+0.21+0.33)$$
$$Y = 3.04$$

This value is equivalent to a grade of G. Hence, the prototype and product grades are the same.

Alternative method

Another way of finding a grade for X is by a similarity comparison by the min/max method between X and the VJB by columns

First column: min/max(X,VJB) = (0.18+0.28+0.26+0.21+0)/(0.26+0.3+0.39+0.34+0.33)
= 0.574

Comparisons with the other columns by the same method leads to the following results

Column	Similarity
E	0.574
V	0.597
G	**0.699**
F	0.327
P	0.679

By inspection of the above results it may be noted that the greatest similarity is with G, thus confirming the previous conclusion.

Similar treatments to the above may be carried out for a product which takes the form of a service rather than an artefact.

7.7 Future outcomes

Future outcomes in evolving systems are the effects of the interaction of actions (or strategies), with different environmental states. The term 'environment' here embraces a wide variety of meanings, both physical and conceptual, including technological, economic, political, scientific and medical types, for example. There are therefore the following two types of independent initial factors:

i) Controlled factors, which are the alternative actions or strategies adopted by the decision makers.
ii) Uncontrolled factors, which are the environmental states that provide the theatre for the actions or strategies. These are outside the influence of the decision makers.

The consequences, or merging, of these two sets of factors results in the system future outcomes, which are the dependent qualities and are measured in various quantitative terms, such as monetary values, product volumes or energy units for example on the one hand, or in qualitative terms of satisfaction on the other. In all cases the outcome is metric is some benefit or utility scale, and a criterion is needed to judge the most favourable outcome amongst the several possibilities. Two other factors are therefore required to complete the specification of a problem and its solution, in addition to the two listed above, these are:

iii) The utilities,or benefits, associated with the various outcomes.
iv) The criterion for selecting the most beneficial appropriate outcome. This may be the maximum value of the outcome, for example. Alternatively, it may be the best of the worst to minimise regret.

Some types of outcome, such as monetary values, are relatively easy to calculate, but others such as those determined by value judgements, are more equivocal. There are also cases where insufficient data is available, such as when the data is difficult to collect or naturally sparse. Of particular interest here is the case where some or all of the data is vague or ambiguous.

Future judgements, the so-called decision making process, may in general terms take place under three different circumstances:

i) Forecasting under certainty (FC). All probabilities are close to unity.
ii) Forecasting under uncertainty (FU). No probabilities are known.
iii) Forecasting under risk (FR). Some or all of the probabilities lie between zero and unity.

The FC case arises when the course of events is highly predictable, because there is a high level of knowledge about closely similar past states and actions. The FU case arises as a result of the potential effects of natural disasters, epidemics, political unrest, legal action or unknown scientific effects, for example. The third case, that of FR, is the focus of interest in the following discussion. In this case there is some knowledge of uncontrolled factors, but it is imperfect and these uncontrolled factors may have a profound influence on the resulting outcomes.

Decision trees and arrays

In the analysis of future outcomes use is frequently made of so-called decision trees, which illustrate the interaction between the environmental states and strategies of a system. The same information may also be displayed in tabular form.

Consider, for example, a simple case of two possible actions, $a_i(i = 1,2)$, associated with two possible states, s_j $(j = 1,2)$. The decision tree starts with a decision point, indicated by the square in Figure 7.1, where either of the a_i actions may be chosen. The two circles are called chance points, since they are associated with uncontrolled factors, which are the s_j and have probabilities p_k $(k = 1,2)$ respectively.

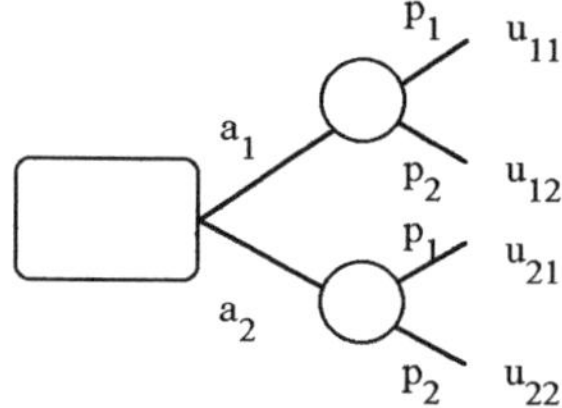

Figure 7.1 A Two-Action, Two-Probability Decision Tree

The information in Figure 7.1 is displayed below in equivalent tabular form, Table 7.5. (Note the transposition of the matrix).

Table 7.5 A Two-Action, Two-Probability Table

States	S_1	S_2	Expected Value
Probability	p_1	p_2	
Actions a_1	u_{11}	u_{21}	v_1
a_2	u_{12}	u_{22}	v_2

There are four benefits, called utilities, u_{ij},one being associated with each (a_i,s_j) pair. They are the outcomes for the FU case. The utilities, u_{ij}, may be computed values or may be subjective expert opinions.

The expected values, v, in Figure 7.1 and Table 7.5 are obtained by matrix multiplication

$$|v_1 \quad v_2| = |p_1 \quad p_2| \begin{vmatrix} u_{11} & u_{21} \\ u_{21} & u_{22} \end{vmatrix} \qquad 7.26)$$

where the transposition of the u_{ij} matrix is maintained.

Thus, $$v_1 = p_1u_{11}+p_2u_{21} \qquad 7.27)$$

and $$v_2 = p_1u_{12}+p_2u_{22} \qquad 7.28)$$

(Reference may be made to any standard text on matrix algebra for further information).

It is a necessary condition that

$$\Sigma_{i=1}{}^{n}p_i = 1 \qquad 7.29)$$

in all cases.

Note that the v values are those that would be obtained on average if the a actions were to be repeated a large number of times.

Example 7.8

Consider a system in which there are three possible environmental states, s_i (i = 1,2,3), and three possible actions, a_j (j = 1,2,3). The probabilities p_k (k = 1,2,3) are; (0.3 0.5 0.20 and the utilities array is

		s_1	s_2	s_3
	a_1	18	8	14
u_{ij} =	a_2	11	7	12
	a_3	-2	6	6

Assume that the optimum outcome criterion is max{v}. Find the corresponding best course of action.

Solution

From the given data, the decision tree may be constructed as shown in Figure Ex. 7.6.

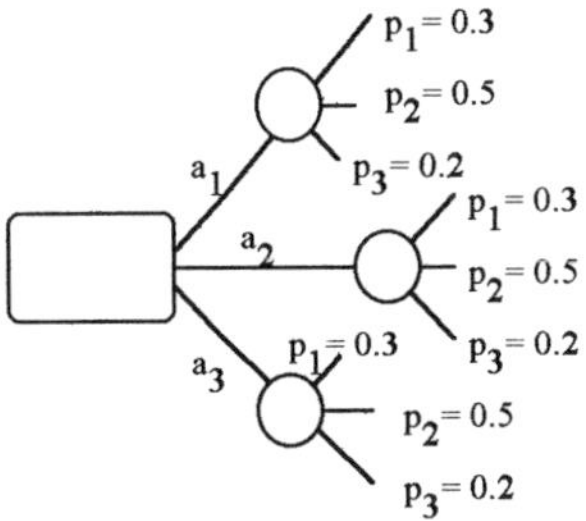

Figure Ex. 7.6 Decision Tree

By matrix multiplication the expected value vector is

$$|v_1\ v_2\ v_3| = |0.3\ 0.5\ 0.2|\begin{vmatrix}18 & 11 & 2\\ 8 & 7 & 6\\ 14 & 12 & 6\end{vmatrix}$$

$$= |12.2\ 0.2\ 3.6|$$

Hence, $\max|v_i| = 12.2$

According to the given maximum expected value criterion, the best course of action is judged to be a_1. The meaning of this is that by choosing action a_1, the actual outcome may be 18,8 or 14 each time the action is repeated. However, if the same action is repeated a large number of times then the average value of the v_i would be 12.2, which is higher than the average v_i that would be obtained were the repeated action either a_2 or a_3. Clearly, if the repeated action is over a series of time intervals then the present values of the utilities arrays would be appropriate for the monetary utilities.

7.8 Fuzzy actions and fuzzy states

The previous section has introduced some concepts of statistics in conventional decision theory in which the states and actions are all non-fuzzy. This foundation can now be extended to the important case where the states and actions are both fuzzy.

To fuzzify the environmental states; consider a universe of discourse of environmental states, s_i (i = 1,2......n) and let this universe of discourse be partitioned into fuzzy sets, S_j (j = 1,2.......m) (m<n). With the partitioning there will be an associated mxn array of membership values,μ_{ij}, which define the S_j fuzzy sets. Also, let the probabilities associated with the s_i fuzzy sets be p_i. Then the probabilities associated with the fuzzy environmental state sets, S_j will be defined by

$$P_j = \Sigma_{i=1}{}^{n}\mu_{ij}p_i \qquad 7.30)$$

Likewise, the universe of discourse of possible actions, a_i, (1 = 1,2.......r) can be fuzzified into fuzzy sets, A_j (j = 1,2......t) (t<r). There will also be an associated rxt array of membership values,ν_{ij}, which define the A_j fuzzy sets.

The result of the interaction between the fuzzy actions A and the fuzzy states S, is governed by the utlilities as shown in Table 7.6.

Table 7.6 Utilities Relating Fuzzy States and Fuzzy Actions

	Fuzzy Actions A_1	A_2............A_t
Fuzzy States		
S_1	U_{11}	U_{12}..........U_{1t}
S_2	U_{21}	U_{22}..........U_{2t}
.	.	. .
.	.	. .
S_m	.U_{m1}	U_{m2}.........U_{mt}

Note that in the double subscripts for the utilities the first subscript refers to the state and the second to the action. The expected values vector of the outcomes is obtained by matrix multiplication of the state set probabilities, equation 7.30), with the utilities matrix U_{ij}, thus

$$|V_1 \; V_2......V_t| = |P_1 \; P_2... \; ....P_m| \begin{vmatrix} U_{11} & U_{12}......U_{1m} \\ U_{21} & U_{22}......U_{2t} \\ \multicolumn{2}{c}{......................} \\ U_{m1} & U_{m2}.....U_{mt}. \end{vmatrix} \qquad 7.31)$$

Example 7.9

The membership values for the fuzzy environmental states, S_1, S_2 and S_3 are shown in Table Ex. 7.8 below, together with the initial probabilities of the s_i states (i = 1,2.......5). The utility array relating to the fuzzy states and actions is given in Table Ex. 7.9.

Table Ex. 7.8 Membership Values

	s_i	s_1	s_2	s_3	s_4	s_5
	p	0.2	0.3	0.3	0.1	0.1
State						
S_1		1	0.5	0	0	0
S_2		0	0.5	1	0.5	0
S_3		0	0	0	0.5	1

Table Ex. 7.9 Utilities Values

Fuzzy Action	A_1	A_2	A_3
State			
S_1	15	9	0
S_2	6	5	3
S_3	12	10	3

Find the maximum expected value.

Solution

From equation 7.30), the probabilities of the S fuzzy states is given by matrix multiplication

$$\begin{vmatrix} P_1 \\ P_2 \\ P_3 \end{vmatrix} = \begin{vmatrix} \mu_{11} & \mu_{12} & \mu_{13} & \mu_{14} & \mu_{15} \\ \mu_{21} & \mu_{22} & \mu_{23} & \mu_{24} & \mu_{25} \\ \mu_{31} & \mu_{32} & \mu_{33} & \mu_{34} & \mu_{35} \end{vmatrix} \begin{vmatrix} p_1 \\ p_2 \\ p_3 \\ p_4 \\ p_5 \end{vmatrix}$$

$$= \begin{vmatrix} 1 & 0.5 & 0 & 0 & 0 \\ 0 & 0.5 & 1 & 0.5 & 0 \\ 0 & 0 & 0 & 0.5 & 0 \end{vmatrix} \begin{vmatrix} 0.2 \\ 0.3 \\ 0.3 \\ 0.1 \\ 0.1 \end{vmatrix}$$

$$= \begin{vmatrix} 0.35 \\ 0.50 \\ 0.15 \end{vmatrix}$$

Now taking the matrix product of the transpose of the fuzzy state set probabilities, (P), with the given utility matrix yields the expected value vector

$$|V_1 \quad V_2 \quad V_3| = |P_1 \; P_2 \quad P_3| \begin{vmatrix} U_{11} & U_{12} & U_{13} \\ U_{21} & U_{22} & U_{23} \\ U_{31} & U_{32} & U_{33} \end{vmatrix}$$

$$= |0.35 \quad 0.5 \;\; 0.15| \begin{vmatrix} 15 & 9 & 0 \\ 6 & 5 & 3 \\ 12 & 10 & 3 \end{vmatrix}$$

$$= |10.05 \quad 7.15 \quad 1.95|$$

The maximum component of the expected value is 10.05, which corresponds with action A_1.

The partitioning of the universe of discourse of the fuzzy action has not been explicitly defined in terms of membership values, but its effect is implicit in the utility matrix. In practice, the partitioning of both the state and action universes of discourse would be effected before the utility matrix components are obtained.

7.9 Updating probabilities

It may happen that there is a management view about the environmental state probabilities, but additional information becomes available which may influence the view of the probabilities. Such may be the case, for example when new economic data is received and investments are being considered. The prior probabilities will therefore need to be reviewed and possibly amended in the light of the new information. There is a well known procedure, called the Bayesian method, by which probabilities may be updated. The method basically involves the calculation of scaling factors derived from the new information, which are used to modify the prior probabilities.

Normally, management would consult several sources of information for new indications of probabilities. Suppose that there are three sources of information (a,b,c) and two possible states, then this would provide six new probabilities

States	a	b	c
s_1	p_{11}'	p_{12}'	p_{13}'
s_2	p_{21}'	p_{22}'	p_{23}'

where the probabilities (p_{ij}') are normalised by rows.

Let there be two prior probabilities, p_1 and p_2, a probability modulus vector, p_i'', is defined by

$$|p_1'' \; p_2'' \; p_3''| = |p_1 \; p_2| \begin{vmatrix} p_{11}' & p_{12}' & p_{13}' \\ p_{21}' & p_{22}' & p_{23}' \end{vmatrix} \qquad 7.32)$$

Note that the sum of the p_i'' is unity.

The array of scaling factors (λ_{ij}) is also defined as a product of probabilities, thus

$$\begin{vmatrix} \lambda_{11} & \lambda_{12} & \lambda_{13} \\ \lambda_{21} & \lambda_{22} & \lambda_{23} \end{vmatrix} = \begin{vmatrix} p_{11}' & p_{12}' & p_{13}' \\ p_{21}' & p_{22}' & p_{23}' \end{vmatrix} \begin{vmatrix} 1/p_1'' & 0 & 0 \\ 0 & 1/p_2'' & 0 \\ 0 & 0 & 1/p_3'' \end{vmatrix} \qquad 7.33)$$

The new updated probabilities (q_{ij}) are now obtained by scaling the prior probabilities with the scaling factors given in equation 7.33)

$$\begin{vmatrix} q_{11} & q_{12} & q_{13} \\ q_{21} & q_{22} & q_{23} \end{vmatrix} = \begin{vmatrix} p_1 & 0 \\ 0 & p_2 \end{vmatrix} \begin{vmatrix} \lambda_{11} & \lambda_{12} & \lambda_{13} \\ \lambda_{21} & \lambda_{22} & \lambda_{23} \end{vmatrix} \quad 7.34)$$

The expected value (v) is obtained by matrix multiplication of the utility matrix with the updated probability matrix

$$\begin{vmatrix} v_{11} & v_{12} & v_{13} \\ v_{21} & v_{22} & v_{23} \end{vmatrix} = \begin{vmatrix} u_{11} & u_{21} \\ u_{12} & u_{22} \end{vmatrix} \begin{vmatrix} q_{11} & q_{12} & q_{13} \\ q_{21} & q_{22} & q_{23} \end{vmatrix} \quad 7.35)$$

Selecting the maximum value from each of the three v columns and expressing these as a vector, the overall expected value is defined by

$$v' = \begin{vmatrix} v_1 & v_2 & v_3 \end{vmatrix} \begin{vmatrix} p_1 \\ p_2 \\ p_3 \end{vmatrix} \quad 7.36)$$

It may be noted that in equation 7.34); $q_{11}+q_{21} = 1$, $q_{12}+q_{22} = 1$ and $q_{13}+q_{23} = 1$.

The method may be applied to fuzzy systems, but the matter is not developed here, it would take the discussion outside the boundaries of an introductory text.

Appendix

A Guide to Fuzzy Logic

Preliminary comments and definitions have been given in Chapter 1. Here the notation is extended and additional operations and concepts are described to provide a reference for the underlying FL treatments used throughout this text.

A1 Notation

The symbolic elements of a set are written in lower case and enclosed in brackets: X= {a,b,c......}, where X represents the set and a,b,c...are the elements of the set.

The so-called Zadeh notation extends the representation of a set to include both the elements of a set and their corresponding degree of membership (μ), which are always in the range of zero to unity. The elements are enclosed in square brackets in this case, thus

$$Z = [\mu_a//a+\mu_b//b+\mu_c//c+......] \quad A1)$$

In expression A1) the double oblique line separates the element and its membership degree and does not represent division. Quantities on the left-hand side of the symbol, (which are always in the range, 0 to 1) are only subject to logic operations, whilst quantities on the right-hand side are only subject to mathematical operations. Also the plus sign means continuation, not addition.

The right-hand side of expression A1) is a collection of discrete elements and their membership values. In the case where the elements form a continuum, say x, the degree of membership is a continuous function of x, $\mu = \mu(x)$. The Zadeh notation in this case is represented by

$$Y = \int \mu(x)//x \quad A2)$$

In this expression, the integral sign represents continuous association.

It is some times convenient to use the discrete form, A1), as an approximation to the continuous distribution, A2). This is frequently the case throughout this text as an aid to the manipulation of thc cxpressions.

Throughout this text, * means algebraic product, but $\times$ means Cartesian product, (see Section A3).

In membership function notation the following symbols are used for one or two universes of discourse (that is, one or two dimensions)

Intersection, $\mu_A(x)\wedge\mu_B(x)$ or $\mu_A(x,y)\wedge\mu_B(x,y)$

$$= \min(\mu_A(x),\mu_B(x)) \text{ or } \min(\mu_A(x,y),\mu_B(x,y))$$

Union, $\mu_A(x)\vee\mu_B(x)$ or $\mu_A(x,y)\vee\mu_B(x,y)$

$$= \max(\mu_A(x),\mu_B(x)) \text{ or } \max(\mu_A(x,y),\mu_B(x,y))$$

It may be noted that the Cartesian product appears a similar operation to the one universe of discourse intersection operation, but conducted between two different universes of discourse to produce a two-dimensional relationship. Also, the Cartesian operator is effected with μ_A as a column vector and μ_B as a row vector, but the min or max operator is effected with μ_A as a row vector and μ_B as a column vector.

A2 Defuzzification

The result of FL operations with fuzzy sets is invariably a conclusion in the form of a fuzzy set. In practice it may sometimes be required to elicit a single finite numerical output, for example, as an output signal in control engineering. This can be achieved by defuzzifying the fuzzy set. There are a variety of methods for rendering a fuzzy set into a single representative value. The two found in this text are the most commonly used methods. If the fuzzy set is in discrete form, expression A1), it may be defuzzified by the formula

$$Z_c = (\mu_a a+\mu_b b+\mu_c c+........)/(\mu_a+\mu_b+\mu_c+......) \qquad \text{A3)}$$

In this expression + means addition.

If the fuzzy set is in continuous form, expression A2), it may be defuzzified by the formula

$$Y_c = \int_x x\mu(x)dx/\int_x(x)dx \qquad \text{A4)}$$

In this expression $\int_x$ means integration over the universe of discourse (x).

A3 Cartesian product

A Cartesian product operation upon two discrete fuzzy sets produces a membership function relational array. The array represents an implication about a conclusion (output) from a given antecedent (input). The Cartesian product is found by comparing pairs of membership values of two fuzzy sets and selecting the minimum value of each pair. Let A and B be the two fuzzy sets and R be the Cartesian product, then the operation is given by

$$R = A\times B \qquad \text{A5)}$$

The membership value of the Cartesian product is given by

$$\mu_R = \min(\mu_A(a),\mu_B(b)) \qquad \text{A6)}$$

For example, let two fuzzy sets be given by

$$A = [0.1//a_1+0.3//a_2+0.7//a_3] \quad \text{and } B = [0.5//b_1+0.8//b_2]$$

Then, $$\mu_R = \min((0.1,0.5)(0.1,0.8)(0.3,0.5)(0.3,0.8)(0.7,0.5)(0.7,0.8)) = (0.1,0.1,0.3,0.3,0.5,0.7) \qquad \text{A7)}$$

This may be cast in the form of an array

R =	a_1	a_2	a_3
b_1	0.1	0.3	0.5
b_2	0.1	0.3	0.7

Each membership value entry in the above tabulation is associated with an (a,b) pair. Alternative relational relays may be formed in other ways, for example

$$\mu_R = \max(\min(\mu_A(a),\mu_B(b)),(1-\mu_A(a))) \qquad \text{A8)}$$

Under certain restricted conditions, expressionA6) and A8) give the same results.

There are several other alternative formulations of relational arrays, but the above two are those used in this text.

A4 Composition

The array, R, formed by the Cartesian product operation, or one of its alternatives, relates, or maps, the elements of universe of discourse A to the elements of the universe of discourse B. Now if the elements of the universe of discourse B are related to those of the universe of discourse C by another array, S, it is possible to find an array which relates the elements of universe of discourse A directly onto those of the universe of discourse C. This latter array, say T, is found by composition

$$T = R \circ S \qquad \text{A9)}$$

where the symbol o denotes fuzzy max-min composition. This may be expressed in membership function notation as

$$\mu_T = \max(\min(\mu_R,\mu_S)) \qquad \text{A10)}$$

The composition operation may be implemented in a similar way to matrix multiplication. For example, let

R =	b_1	b_2
a_1	0.6	0.4
a_2	0.7	0.3

S =	c_1	c_2	c_3
b_1	0.2	0.2	0.5
b_2	0.1	0.3	0.7

Now, let $\quad T = R \circ S$

$$= \text{maxmin} \begin{vmatrix} 0.6 & 0.4 \\ 0.7 & 0.3 \end{vmatrix} \begin{vmatrix} 0.2 & 0.2 & 0.5 \\ 0.1 & 0.3 & 0.7 \end{vmatrix}$$

$$= \max \begin{vmatrix} (0.2,0.1)(0.2,0.3)(0.5,0.4) \\ (0.2,0.1)(0.4,0.3)(0.5,0.3) \end{vmatrix}$$

Hence, $T = \begin{vmatrix} 0.2 & 0.3 & 0.5 \\ 0.2 & 0.4 & 0.5 \end{vmatrix}$

A5 Extension principle

This principle enables mathematical functions of fuzzy sets to be defined and evaluated. The principle states that if a mathematical function is applied to a fuzzy set, then it should be applied to each of the fuzzy set elements.

Let there be a fuzzy set A, and let there be a function of the set f(A), where

$$A = [\mu_1//a_1+\mu_2//a_2+\mu_3//a_3+........]$$

Then $$f(A) = f[\mu_1//a_1+\mu_2//a_2+\mu_3//a_3+........] \quad \text{A11)}$$

By the extension principle

$$f(A) = [\mu_1//f(a_1)+\mu_2//f(a_2)+\mu_3//f(a_3)+........] \quad \text{A12)}$$

For example, let f(a) = a^2 and let there be a number A "about 4"

$$A = [0.1//2+0.5//3+1.0//4+0.5//5+0//6]$$

Then, $$f(A) = [0.1//4+0.5//9+1.0//16+0.5//25+0//36]$$

As another example, let B represent another fuzzy number "about 5". The algebraic product (C) of the fuzzy numbers A and B is

$$C = A*B \quad \text{A13)}$$

If, $$B = [0.5//4+1.0//5+0.5//6]$$

Then, $$C = [0.1//2+0.5//3+1.0//4+0.5//5+0//6]*[0.5//4+1.0//5+0.5//6]$$

$$= [\min(0.1,0.5)//8+\min(0.1,1.0)//10+\min(0.1,0.5)//12 +\min(0.5,0.5)//12+\min(0.5,1.0)//15+............]$$

$$= [0.1//8+0.1//10+0.1//12+0.5//12+0.5//15+0.5//18+0.5//16 +0.5//20+0.5//24+0.5//20+0.5//25+0.5//30+0//24+0//30+0//36]$$

Where entries in this expression are duplicated, then one may be eliminated giving,

$$C = [0.1//8+0.1//10+0.1//12+0.5//12+0.5//15+0.5//16+0.5//18+0.5//20 +0.5//24+0.5//25+0.5//30+0//30]$$

A6 Principal sets

The result of mathematical operations on discrete fuzzy sets may usually be expressed in the form of an array. By inspection of the array, the minimum and maximum values of the elements in the array for each membership level may be ascertained. The collection of these minimum and maximum elements and associated membership values constitute the principal fuzzy set, which has the property of containing within its envelope all other elements. This results in considerable simplification where manipulation of the sets is required.

Consider for example the fuzzy set C above, the elements at the various membership levels are listed below

Membership level	**Elements**		
0.1	**8**	10	**12**
0.5	**12**	15	18
0.5	16	20	24
0.5	20	25	**30**
0	24	**30**	36

The principal fuzzy set is shown in bold type in the above array and illustrated in Figure A1 below. The remainder of the elements, including 0//24 are subsumed within the envelope of the principal set

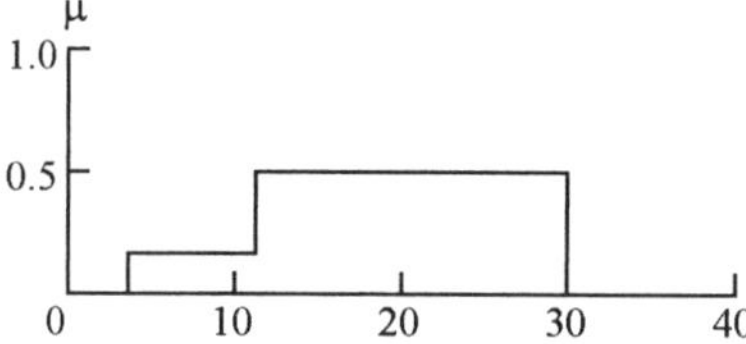

Figure A1 Principal Fuzzy Set C

The fuzzy set shown in Figure A1 in which all membership values are less than unity is known as a sub-normal set.

As another example, consider two fuzzy sets, X and Y

where $X = [0//0.5+0.5//1.0+0.7//1.5+1.0//2.0+0.7//2.5+0.5//3.0+0//3.5]$
$Y = [0//0.5+0.5//1.0+1.0//1.5+0.5//2.0+0//2.5]$

The algebraic product (Z) of X and Y is

$$Z = X*Y$$

There results two arrays, one of which represents the Cartesian product of the membership values, whilst the other represents the arithmetic product of the elements. These arrays are shown in Table A1 and Table A2.

Table A1 Membership Values Array

y \ x	0.5	1.0	1.5	2.0	2.5	3.0	3.5
0.5	**0**	0	0	0	0	0	0
1	0	**0.5**	0.5	0.5	0.5	0.5	0
1.5	0	0.5	**0.7**	**1.0**	**0.7**	0.5	0
2	0	0.5	0.5	0.5	0.5	**0.5**	0
2.5	0	0	0	0	0	0	**0**

Table A2 Arithmetic Product of Elements Array

y \ x	0.5	1.0	1.5	2.0	2.5	3.0	3.5
0.5	**0.25**	0.50	0.75	1.00	1.25	1.50	1.75
1	0.50	**1.00**	1.50	2.00	2.50	3.00	3.50
1.5	0.75	1.50	**2.25**	**3.00**	**3.75**	4.50	5.25
2	1.00	2.00	3.00	4.00	5.00	**6.00**	7.00
2.5	1.25	2.50	3.75	5.00	6.25	7.50	**8.75**

The principal fuzzy set is shown in bold type in the above Tables and is illustrated in Figure A2.

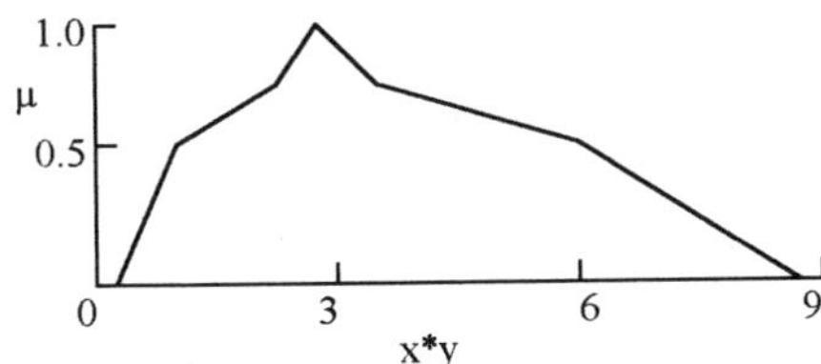

Figure A2 Fuzzy Set Z

A7 Grouping by similitude

Grouping (clustering or classification) is a basic scientific activity with the objective of creating general concepts and it is therefore of considerable importance. There are many methods of grouping, each depending upon the definition of a metric, the identification of a feature of the objects (real or virtual) and an optimising principle.

The technological applications of this are widespread in such areas as pattern recognition, system identification and clustering in group technology.

The method described here is called the min-max method and it is applicable to a quite wide range of technological problems, it is also one of the simplest to apply. If the data to be compared is in tabulated form, rows are compared with each other and similarities are found in the range,0 to 1, the most similar pair would naturally produce the fraction closest to unity. A similar analysis is performed on the columns. Such an analysis enables the rows and columns to be grouped together at various levels of similarity until all the rows and columns are exhausted.

Consider an array of elements as shown in Table A3:

Table A3 Tabulated Data

	X_1	X_2	X_3...etc	X_n
Y_1	a_{11}	a_{12}	a_{13}..etc	a_{1n}
Y_2	a_{21}	a_{22}	a_{23}..etc	a_{2n}
.	.	.	.	.
.	.	.	.	.
Y_m	.	.	.	a_{mn}

The column entries are compared by the min-max method which defines a metric r_{ij}, thus

$$r_{ij} = \Sigma\min(a_{ki},a_{ki})/\Sigma\max(a_{ki},a_{kj}) \qquad \text{A14)}$$

or

$$= (\min(a_{1i},a_{1j})+\min(a_{2i},a_{2j})+........\min(a_{mi},a_{mj}))/ (\max(a_{1i},a_{1j})+\max(a_{2i},a_{2j})+...\max(a_{mi},a_{mj})) \qquad \text{A15)}$$

The r_{ij} values are always in the range 0 to 1 and they compare the similarity of the two columns. An mxn array is formed by the r_{ij}. This will necessarily be reflexive and symmetrical, that is $r_{ii} = 1$ and $r_{ij} = r_{ji}$. A similarity analysis of the rows can be achieved by an identical method.

Example A1

Group the columns and rows of the binary entries shown in Table Ex. A1.

Table Ex. A1 Sample Tabulated Binary Data

				x				
	1	2	3	4	5	6	7	8
y								
1	1	1			1			
2				1				1
3		1	1			1	1	
4				1				1
5			1	1		1	1	
6	1	1			1			

A typical operation would be the comparison of columns X_2 and X_5, thus

$$r_{25} = (\min(1,1)+\min(1,0)+\min(1,1))/(\max(1,1)+\max(1,0)+\max(1,1)) = 2/3$$

The remaining r are evaluated in the same way to produce the symmetrical similarity array, C, shown in Table Ex. A2.

Table Ex. A2 Similitude of Data by Columns

$$C = \begin{array}{cccccccc} 1 & & & & & & & \\ 2/3 & 1 & & & & \text{Sym.} & & \\ 0 & 1/4 & 1 & & & & & \\ 0 & 0 & 1/4 & 1 & & & & \\ 1 & 2/3 & 0 & 0 & 1 & & & \\ 0 & 1/4 & 1 & 1/4 & 0 & 1 & & \\ 0 & 1/4 & 1 & 1/4 & 0 & 1 & 1 & \\ 0 & 0 & 0 & 2/3 & 0 & 0 & 0 & 1 \end{array}$$

Rows of the binary table may be compared in the same way producing another symmetrical similarity array, R, shown in Table Ex. A3.

Table Ex. A3 Similitude of Data by Rows

$$R = \begin{array}{cccccc} 1 & & & & & \\ 0 & 1 & & & & \\ 1/6 & 0 & 1 & \text{Sym.} & & \\ 0 & 1 & 0 & 1 & & \\ 0 & 1/5 & 3/5 & 1/5 & 1 & \\ 1 & 0 & 1/6 & 0 & 0 & 1 \end{array}$$

By inspection of Table Ex. A2 the following list of column similarity rankings in decreasing order may be obtained

(1,5)(3,6)(3,7) = 1
(4,8)(2) = 2/3

At the 2/3 level, the list of entries is exhausted.

Since 3 appears twice in the unity ranking, the groups may be condensed to

(1,5)(3,6,7) = 1

Also, by inspection of Table Ex. A3 the following row similarity rankings in decreasing order may be obtained

(1,6)(2,4) = 1
(3,5) = 3/5

At the 3/5 level, the list of entries is exhausted.

In the similarity rankings, 1 represents identity, whilst 0 represents complete non-similarity.

There are a number of other similarity methods published in the literature; cosine amplitude, geometric average minimum and arithmetic average minimum, to name a few. Equivalent results are obtained to those given above, but generally with increased computation, and are therefore not discussed here.

It will be noted that the similitude method is not restricted to binary data.

A8 Grouping by composition

As remarked before, the result of a min-max operation on given data is to produce an array that is both reflexive and symmetrical, but not necessarily transitive, that is, it does not necessarily satisfy the equivalence condition. For example, let Table A4 be a given 7x5 binary data array.

Table A4 7x5 Binary Data Array

y \ x	1	2	3	4	5	6	7
1	1	1	1		1		
2			1	1	1		
3	1			1		1	
4				1		1	
5		1	1				1

By the min-max method, the similitude array of rows is obtained as shown in Table A5:

Table A5 Similitude of Rows

	y \ x	1	2	3	4	5
	1	1				
	2	2/5	1	Sym.		
R =	3	1/6	1/5	1		
	4	0	1/4	2/3	1	
	5	2/5	1/5	0	0	1

Table A6 The R^2 Array

	y \ x	1	2	3	4	5
	1	1				
	2	2/5	1	Sym.		
R^2 =	3	1/5	1/4	1		
	4	1/4	1/4	2/3	1	
	5	2/5	2/5	1/5	1/5	1

Table A6 shows the R^2 array, which is obtained from the R array, Table A5, by composition, as described in Section A4 above.

Now consider the entries in Table A5 and Table A6 for the entries at locations (1,2),(2,4) and (1,4). The entries are listed below for reference

Location	R	R^2
(1,2)	2/5	2/5
(2,4)	1/4	1/4
(1,4)	0	1/4

It may be noted that the R table entries fail the test

$$r_{14} \geq \min(r_{12}, r_{24})$$

But the R^2 entries satisfy the condition. The R^2 array satisfy the equivalence condition (reflexive,symmetrical and transitive) whereas the R array satisfies only the tolerance condition (reflexive and symmetrical). If it happened that the R^2 array fails to satisfy the equivalence condition, then further composition operations on the array may be performed until the condition is achieved. (At the most (n-1) compositions are required, where n is the cardinality of the array).

The value of the equivalence condition in this context is that it enhances the definition of groups in the array. For example, in Table A6, let all the entries with $r_{ij} \geq 2/5$ be identified, the result is shown in Table A7 below.

Table A7 The R^2 Array for $r_{ij} \geq 2/5$

	1	2	3	4	5
1	1	1			1
2	1	1			1
3			1	1	
4			1	1	
5	1	1			1

It is immediately clear by inspecting Table A7 that columns (or rows) (1,2,5) are identical and also that columns (or rows) (3,4) are also identical. In the case of a data array with few entries, as in the above example, the additional work required to find equivalence may not be justified, but for a large array it can be beneficial.

A.9 Inter-group relations

As will be apparent from the foregoing discussion, it is always possible to form two sets of groups from any data array, one set each for similitude by rows and columns. It is then possible to develop this further by considering the relationships between the two sets of groups so formed.

Consider again the binary data array shown in Table A4. Through a similitude analysis by columns a second set of groups is found: (2,3,5),(4,6),(1) and (7). The two sets of groups are linked below in Figure A3.

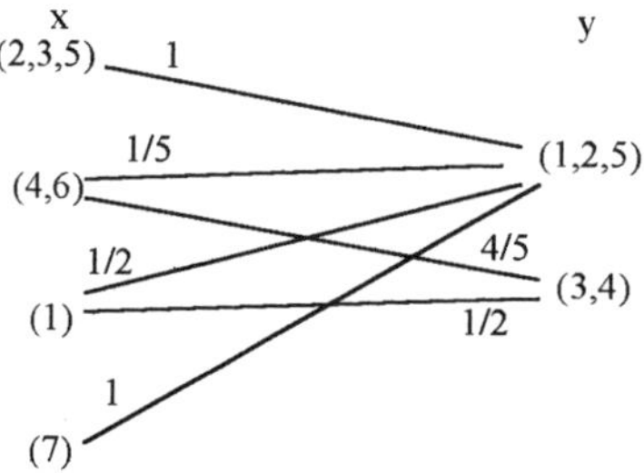

Figure A3 Relationship Strengths Between X and Y Groups

By inspection of the data in Table A4, the proportion of entries linking each pair of (x,y) groups may be found. Thus out of the five entries for the (4,6) x-group, there are four allocated to (3,4) y-group and one to the (1,2,5) y-group. Each link in Figure A3 may thus be allocated a "strength" as shown.

A.10 Multiple antecedents

A first-order FL proposition has one antecedent, it is of the form

IF A THEN X

The associated rule-base relating the antecedent and consequent fuzzy sets is linear in form

A	A_1	A_2...etc	A_n
X	X_1	X_2...etc	X_n

A second-order FL proposition has two antecedents and is of the form

IF A AND B THEN X

The associated rule-base is two-dimensional in form

	A_1	A_2....etc	A_n
B_1	X_{11}	X_{12}..etc	X_{1n}
B_2	X_{21}	X_{22}..etc	X_{2n}
.	.	....	
B_m	X_{m1}	X_{m2}...etc	X_{mn}

In a similar way, a third-order proposition would require a three-dimensional rule-base array. Beyond that, a hyper-space rule-base representation would be required.

A higher order rule-base can however be reduced to a lower order and if the process is repeated a sufficient number of times then a rule-base of any order may be linearised. Consider for example a second-order rule base

IF A AND B THEN X = IF C THEN X

where C = (A AND B).

The second-order form is therefore formally reduced to a linear form.

Example A2

Consider a 3x3 rule-base array and reduce it to a linear form.

Let the rule-base be represented by

	A_1	A_2	A_3
B_1	X_{11}	X_{12}	X_{13}
B_2	X_{21}	X_{22}	X_{23}
B_3	X_{31}	X_{32}	X_{33}

By formal transformation, this may be represented by

A_1' A_2'...etc A_9'
V_1 V_2.....etc V_9

where $A_1' = A_1B_1$, $A_2' = A_2B_1$ etc and $A_9' = A_3B_3$

also $V_1 = X_{11}$, $V_2 = X_{12}$ etc and $V_9 = X_{33}$

Thus the 3x3 array is transformed into a 1x9 array.

Similarly, a 3x4 array (or a 2x6 array) may be transformed into a 1x12array.

Higher order arrays, such A AND B AND C may be reduced to (A AND B) AND C or A AND (B AND C) or (A AND B AND C). This is a convenient reduction, enabling most practical orders of arrays to be represented in one or two-dimensional space rather than hyperspace.

A.11 Adjectival processes

An adjectival process is one which confers a qualifying meaning on a fuzzy set, usually by concentrating or by diluting the set character. For example, if a fuzzy set represents 'LOW', then 'VERY LOW' or 'LOW LOW' is represented by

$$\text{VERY LOW} = (\text{LOW})^2 \qquad \text{A16)}$$

If, for simplicity, 'LOW' is expressed by a singleton

$$\text{LOW} = \mu//a \qquad \text{A17)}$$

Then $(LOW)^2 = \mu^2//a$ A18)

If $\mu = 1$, then the process has no effect, but for $\mu<1$ the membership value is diminished.

Let LOW = [0//a+0.5//b+1.0//c+0.5//d+0//e] A19)

Then, $(LOW)^2 = [0//a+0.5^2//b+1.0^2//c+0.5^2//d+0//e]$ A20)
$= [0//a+0.25//b+1.0//c+0.25//d+0//e]$

The effect of this process is to concentrate the value around 'c', that is, 'c' is strengthened. The more this type of process is continued, the closer the fuzzy set approaches a singleton representation.

Now consider the case of a fuzzy set representing 'SLIGHTLY LOW', this could be interpreted as $(LOW)^{1/2}$. Thus

$$(LOW)^{1/2} = [0//a+0.5^{1/2}//b+1.0^{1/2}//c+0.5^{1/2}//d+0//e] \quad A21)$$
$$= [0//a+0.707//b+1.0//c+0.707//d+0//e]$$

The effect in this case is to dilute the strength of the fuzzy set around 'c'.

There are other recognised implications such as

$$(LOW\ PLUS) = (LOW)^{1.2}$$

and $(LOW\ MINUS) = (LOW)^{0.8}$

These adjectival processes known as "hedges", are useful ways of adjusting fuzzy sets and may be used for example, in adaptive fuzzy control systems.

A 12 Correlation metric

The correlation metric appears in a number of different guises and is essentially a normalised assessment of the degree of overlap between two fuzzy sets. It is defined as follows. Consider two fuzzy sets, A and B, that have the same universe of discourse. Let these two fuzzy sets be expressed by,

$$A = [\mu_a/a+\mu_b/b+..............\mu_r/r] \quad A22)$$

and $B = [\mu_a'/a+\mu_b'/b+..............\mu_r'/r]$ A23)

The correlation metric, α, is defined by

$$\alpha = A\cap B/A\cup B = \min(\mu_i,\mu_i')/\max(\mu_j,\mu_j') \quad A24)$$

This is a useful concept when assessing the degree of fuzziness in a set. In this case let B be interpreted as A', the complement of A. Then the fuzziness metric, β is expressed as

$$\beta = A \cap A'/A \cup A' \tag{A25}$$

For example, let

$$A = [0//a+0.2//b+0.2//c+0.5//d+0.8//e] \tag{A26}$$

and

$$B = [1.0//a+0.5//b+0.2//c+0//d+0//e] \tag{A27}$$

(Note that A and B have the same span on the universe of discourse).

Then, $A \cap B = 0.4$ and $A \cup B = 3.0$. The correlation metric, α is therefore = 0.1333.

Clearly,

$$A' = [1.0//a+0.8//b+0.8//c+0.5//d+0.2//e] \tag{A28}$$

Hence, $A \cap A' = 1.1$ and $A \cup A' = 3.0$. The fuzziness metric, β, is therefore = 0.3667. Similarly, the fuzziness metric of fuzzy set B is 0.1489. It may be noted that the correlation metric defined above may also be applied to an array.

FURTHER READING

The following are entry point references, each containing a list of references to further aspects in the topics.

A general review of fuzzy logic theory and applications is given in "Fuzzy Logic with Engineering Applications" by T.J. Ross, published by McGraw-Hill Inc., U.S.A., 1995. ISBN 0 07 1136371. A collection of essays of a more theoretical nature is found in "Fuzzy Engineering" edited by B. Kosco, published by Prentice-Hall Inc., U.S.A., 1997. ISBN 0 13 353731 5.

Reference may also be made to the following:

Chapter 1.
Andriole, S.J., "Information Systems Engineering", in Encyclopedia of Science and Technology, Vol. 10, 8th. Ed., pp 163-165, publ. by McGraw-Hill Inc., U.S.A., 1997. ISBN 0 07 911504 7.
Pham, D.T. and Pham, P.T.N., "Artificial Intelligence in Engineering", Intn. Jour. of Mach. Tools and Manufacture, publ. by Pergamon Press, 1999, **39**, pp 937-949.
Pham, D.T. and Dimov, S.S., "An Efficient Algorithm for Automatic Knowledge Acquisition", Pattern Recognition, Publ. by Pergamon Press, 1997, **30**,(7), pp 1137-1143.

Chapter 2.
Perry, J.H., Chilton, C.H. and Kirkpatrick, S.D. (Eds.), "Chemical Engineers Handbook, 7th. Ed., 1997. ISBN 0 0704 98415 (hdbk.) or ISBN 0 0711 54485 (pbk.).
(Also relevant to Chapter 3).

Chapter 3.
Rogers, G.F.C. and Mayhew, Y.R., "Engineering Thermodynamics Work and Heat Transfer", Publ. by Longman Ltd., U.K., 1996. ISBN 0 58 208846 1.

Chapter 4.
Atkins, A.G., "Stress Analysis", Kempes Engineers Year Book edited by Hall Stephens, pp A7/167-A7/224, publ. by Miller Freeman UK Ltd., 1999. ISBN 0 86 382 4080.
Benham, P.P. and Crawford, R.J., "Mechanics of Engineering Materials", publ. by Longman Sci. & Tech., U.K., 1988. ISBN 0 582 025575.
Williams, J.G., "Stress Analysis of Polymers", publ. by Longman Group Ltd., U.K., 1973, ISBN 0582 47001 3.
Thornton, J.N. and Bramley, A.N., "An Approximate Method for Predicting Metal Flow in Forging and Extrusion Operations", Proc. Inst. Mech. Engrs. (U.K.), 1980, **95**,(2), pp 9-15.

Chapter 5.
Reznik, L., "Fuzzy Controllers", publ. by Newnes U.K., 1997, ISBN 0 7506 3429 4.
(Also refers to FLC hardware).
Watts, I. and Cooper, B.J., "Control Systems", in Kempes Engineers Year Book, edited by Hall Stephens, pp A10/305-A10/365, publ. by Miller Freeman UK Ltd., 1999. ISBN 0 86 382 4080.
Minorsky, N., "Theory of non-Linear Control Systems", publ. by McGraw-Hill Inc., U.S.A., 1969.

Chapter 6.
Thomas, M.V., "Industrial Engineering", in Encyclopedia of Science and Engineering, Vol. 10, 8th. Ed., pp 111-113, publ. by McGraw-Hill Inc., U.S.A., 1997. ISBN 0 07 911504 7.
Currie, R.M and Faraday, J.E., "Work Study" 4th.Ed., publ. by Pitman Publishing Ltd., 1977. ISBN 0 273 009591.
Muhlemann, A.,et al.,"Production and Operations Management", 6th. Ed., publ. by Pitman Publishing Ltd., 1992. ISBN 0 273 032356.
Dale, B. and Oakland, J., "Quality Improvement through Standards", 2nd.Ed., publ. by Stanley Thornes Ltd., U.K., 1994. ISBN 0 7487 1699 8.
Greasley, A., "Operations Management in Business", publ. by Stanley Thornes Ltd., U.K., 1999. ISBN 0 7487 2084 7.

Chapter 7.
Mitchell, G., "The Practice of Operational Research", publ. by Macmillan Ltd., U.K., 1992 ISBN 0 4 719 398 2X.
Lesso, W.G., "Operations Research", in Encyclopedia of Science and Engineering, Vol. 10, 8th. Ed., pp 399-407, publ. by McGraw-Hill, U.S.A., 1997. ISBN 0 07 911504 7.

A comprehensive and frequently updated list of the latest research publications in the field may be found on the website; **www.eevl.ac.uk/ram/**, search for Fuzzy Logic. This site currently gives access to more than five hundred relevant abstracts.

Index

International Series on
MICROPROCESSOR-BASED AND INTELLIGENT SYSTEMS ENGINEERING

Editor: Professor S. G. Tzafestas, *National Technical University, Athens, Greece*

1. S.G. Tzafestas (ed.): *Microprocessors in Signal Processing, Measurement and Control.* 1983 ISBN 90-277-1497-5
2. G. Conte and D. Del Corso (eds.): *Multi-Microprocessor Systems for Real-Time Applications.* 1985 ISBN 90-277-2054-1
3. C.J. Georgopoulos: *Interface Fundamentals in Microprocessor-Controlled Systems.* 1985 ISBN 90-277-2127-0
4. N.K. Sinha (ed.): *Microprocessor-Based Control Systems.* 1986 ISBN 90-277-2287-0
5. S.G. Tzafestas and J.K. Pal (eds.): *Real Time Microcomputer Control of Industrial Processes.* 1990 ISBN 0-7923-0779-8
6. S.G. Tzafestas (ed.): *Microprocessors in Robotic and Manufacturing Systems.* 1991 ISBN 0-7923-0780-1
7. N.K. Sinha and G.P. Rao (eds.): *Identification of Continuous-Time Systems.* Methodology and Computer Implementation. 1991 ISBN 0-7923-1336-4
8. G.A. Perdikaris: *Computer Controlled Systems.* Theory and Applications. 1991 ISBN 0-7923-1422-0
9. S.G. Tzafestas (ed.): *Engineering Systems with Intelligence.* Concepts, Tools and Applications. 1991 ISBN 0-7923-1500-6
10. S.G. Tzafestas (ed.): *Robotic Systems.* Advanced Techniques and Applications. 1992 ISBN 0-7923-1749-1
11. S.G. Tzafestas and A.N. Venetsanopoulos (eds.): *Fuzzy Reasoning in Information, Decision and Control Systems.* 1994 ISBN 0-7923-2643-1
12. A.D. Pouliezos and G.S. Stavrakakis: *Real Time Fault Monitoring of Industrial Processes.* 1994 ISBN 0-7923-2737-3
13. S.H. Kim: *Learning and Coordination.* Enhancing Agent Performance through Distributed Decision Making. 1994 ISBN 0-7923-3046-3
14. S.G. Tzafestas and H.B. Verbruggen (eds.): *Artificial Intelligence in Industrial Decision Making, Control and Automation.* 1995 ISBN 0-7923-3320-9
15. Y.-H. Song, A. Johns and R. Aggarwal: *Computational Intelligence Applications to Power Systems.* 1996 ISBN 0-7923-4075-2
16. S.G. Tzafestas (ed.): *Methods and Applications of Intelligent Control.* 1997 ISBN 0-7923-4624-6
17. L.I. Slutski: *Remote Manipulation Systems.* Quality Evaluation and Improvement. 1998 ISBN 0-7932-4822-2
18. S.G. Tzafestas (ed.): *Advances in Intelligent Autonomous Systems.* 1999 ISBN 0-7932-5580-6
19. M. Teshnehlab and K. Watanabe: *Intelligent Control Based on Flexible Neural Networks.* 1999 ISBN 0-7923-5683-7
20. Y.-H. Song (ed.): *Modern Optimisation Techniques in Power Systems.* 1999 ISBN 0-7923-5697-7
21. S.G. Tzafestas (ed.): *Advances in Intelligent Systems.* Concepts, Tools and Applications. 2000 ISBN 0-7923-5966-6
22. S.G. Tzafestas (ed.): *Computational Intelligence in Systems and Control Design and Applications.* 2000 ISBN 0-7923-5993-3
23. J. Harris: *An Introduction to Fuzzy Logic Applications.* 2000 ISBN 0-7923-6325-6